Lecture Notes in Computer Science 9655

Commenced Publication in 1973
Founding and Former Series Editors:
Gerhard Goos, Juris Hartmanis, and Jan van Leeuwen

More information about this series at http://www.springer.com/series/8851

Ngoc Thanh Nguyen · Ryszard Kowalczyk (Eds.)

Transactions on Computational Collective Intelligence XXII

Editor-in-Chief

Ngoc Thanh Nguyen
Wroclaw University of Technology
Wroclaw
Poland

Co-editor-in-Chief

Ryszard Kowalczyk
Swinburne University of Technology
Hawthorn
Australia

ISSN 0302-9743   ISSN 1611-3349 (electronic)
Lecture Notes in Computer Science
ISBN 978-3-662-49618-3   ISBN 978-3-662-49619-0 (eBook)
DOI 10.1007/978-3-662-49619-0

Library of Congress Control Number: 2015955870

Printed on acid-free paper

This Springer imprint is published by SpringerNature
The registered company is Springer-Verlag GmbH Berlin Heidelberg

Transactions on Computational Collective Intelligence XXII

Preface

It is my pleasure to present to you the XXII volume of LNCS *Transactions on Computational Collective Intelligence*. This volume inaugurates year 2016, the sixth year of TCCI activities. In 22 issues we have published 222 high-quality papers. This issue contains 11 papers.

The first paper "Pairwise Comparisons Rating Scale Paradox" by Waldemar W. Koczkodaj is devoted to the solution based on normalization of the paradox of unprocessed rating scale data. The author shows that the pairwise comparisons method is the most amazing and universal approach to assessments and decision-making problems.

The second paper entitled "On Achieving History-Based Move Ordering in Adversarial Board Games Using Adaptive Data Structures" by Spencer Polk and B. John Oommen concerns the problem of enhancing the well-known alpha–beta search technique for intelligent game playing. The authors show that, while using lightweight, efficient ranking techniques associated with an adaptive data structure, the mechanism they proposed is able to obtain substantial gains in tree pruning in both the two-player and multi-player cases, in a variety of games.

In the third paper, "Identification of Possible Attack Attempts Against Web Applications Utilizing Collective Assessment of Suspicious Requests," Marek Zachara presents a new method for detecting attacks against Web applications, in which cooperating systems analyze incoming requests, identify potential threats, and present them to other peers. The method was tested using data from seven different Web servers, consisting of over three million recorded requests.

The fourth paper, "A Grey Approach to Online Social Networks Analysis" by Camelia Delcea et al., presents a model for analyzing whether people from a randomly chosen sample are comparing themselves with the ones in their own network by considering the posts their friends are making on Facebook and whether there is any dependency between the social comparison orientation and the appearance of a negative feeling.

The fifth paper entitled "ReproTizer: A Fully Implemented Software Requirements Prioritization Tool" by Philip Achimugu et al. presents a software named ReproTizer (Requirements Prioritizer), which serves to engender real-time prioritization of software requirements. ReproTizer consists of a weight scale that gives project stakeholders the ability to perceive the influence the different requirements weights may have on the final results.

In the sixth paper, "A Consensus-Based Method for Solving Concept-Level Conict in Ontology Integration," Trung Van Nguyen and Hanh Huu Hoang present a novel method for finding the consensus in ontology integration at the concept level. Their approach is based on the consensus theory and distance functions between attribute values, which gives quite interesting results.

The next paper, "Enhancing Collaborative Filtering Using Implicit Relations in Data," by Manuel Pozo et al. presents a recommender system that relies on distributed recommendation techniques and implicit relations in data. The authors extends matrix factorization techniques by adding implicit relations in an independent layer. Owing to this, they have achieved good results of recommendation process.

In the eighth paper entitled "Semantic Web-Based Social Media Analysis," Liviu-Adrian Cotfas et al. propose a novel semantic social media analysis platform, which is able to properly emphasize users' complex feelings such as happiness, affection, surprise, anger, or sadness.

In the ninth paper, "Web Projects Evaluation Using the Method of Significant Website Assessment Criteria Detection," Paweł Ziemba et al. analyze the applicability of feature selection methods in the task of selecting website assessment criteria to which weights are assigned. The authors tested the applicability of the chosen methods against the approach in which the weightings of website assessment criteria are defined by users. They propose a selection procedure for significant choice criteria and reveal undisclosed user preferences based on the website quality assessment models.

In the tenth paper entitled "Dynamic Database by Inconsistency and Morphogenetic Computing," Xiaolin Xu et al. present a formal description of database transformations in a way to classify the database or to generate a new database from the previously known database. Transformation can be isomorphic or non-isomorphic. Owing to this, the authors have proved that big data can reduce its complexity and be controlled in a better way by its homotopic parts.

The last paper, "A Method for Size and Shape Estimation in Visual Inspection for Grain Quality Control in the Rice Identification Collaborative Environment Multi-agent System," authored by Marcin Hernes et al. presents a method of estimating the size and shape of grain cereals using visual quality analysis. The authors implemented this method in a multi-agent system. They show that using this method should improve the statistical quality of the rice selection and should enable the identification of species/varieties of cereals and determination of the percentage of the grains that do not meet quality standards.

I would like to thank all the authors for their valuable contributions to this issue and all the reviewers for their opinions, which helped maintain the high quality of the papers. My special thanks go to the team at Springer, who helps publish TCCI issues in due time and in good order.

January 2016   Ngoc Thanh Nguyen

Transactions on Computational Collective Intelligence

This Springer journal focuses on research in applications of the computer-based methods of computational collective intelligence (CCI) and their applications in a wide range of fields such as the Semantic Web, social networks, and multi-agent systems. It aims to provide a forum for the presentation of scientific research and technological achievements accomplished by the international community.

The topics addressed by this journal include all solutions to real-life problems for which it is necessary to use CCI technologies to achieve effective results. The emphasis of the papers published is on novel and original research and technological advancements. Special features on specific topics are welcome.

Tadeusz Szuba   AGH University of Science and Technology, Poland
Kristinn R. Thorisson   Reykjavik University, Iceland
Gloria Phillips-Wren   Loyola University Maryland, USA
Sławomir Zadrożny   Institute of Research Systems, PAS, Poland
Bernadetta Maleszka   Wroclaw University of Technology, Poland

Contents

Pairwise Comparisons Rating Scale Paradox

W.W. Koczkodaj[(✉)]

Computer Science, Laurentian University, Sudbury, ON, Canada
`wkoczkodaj@cs.laurentian.ca`

Abstract. This study demonstrates that incorrect data are entered into a pairwise comparisons matrix for processing into weights for the data collected by a rating scale. Unprocessed rating scale data lead to a paradox. A solution to it, based on normalization, is proposed. This is an essential correction for virtually all pairwise comparisons methods using rating scales. The illustration of the relative error, currently taking place in numerous publications, is discussed.

Keywords: Pairwise comparison · Rating scale · Normalization · Inconsistency · Paradox · AHP · Analytic Hierarchy Process

1 Introduction

Thurstone's Law of Comparative Judgments, introduced [14] in 1927 was a milestone in pairwise comparisons (PCs) research although the first documented use of PCs is traced to Ramond Llull in 13th century. A considerable number of customizations, based on different rating scales, have been proposed. Some of them have generated controversies which are not the subject of this study. This study is independent of pairwise comparisons customizations. It concentrates on the theoretical aspects of the rating scale, leaving to the originators of PCs customizations, to accommodate appropriate corrections. The starting point for this study is a PC matrix. Numerous examples demonstrated that the pairwise comparisons can be used to draw conclusions in a comparatively easy and elegant way. The brilliance of the pairwise comparisons could be reduced to a common sense rule: take two at a time if we are unable to handle more than that. For relating one item to another item in a pair, PCs relies on a rating scale "1 to m", where 1 denotes equality and m is used to reflect superiority ("advantage" or some kind of "perfection") of one item above the other item.

In simple language, a rating scale is a set of categories designed to elicit data about a quantitative or a qualitative stimuli (or attribute). It requires a rater to choose a numeric value, sometimes by using a graphical object (e.g., line), to the rated entity, as a measure of some rated stimuli. One of the best known examples of such a scale is a "scale of 1 to 10", or "scale from 1 to 10" where 10 stands for some kind of perfection.

Partially supported by Euro grant Human Capital.

© Springer-Verlag Berlin Heidelberg 2016
N.T. Nguyen and R. Kowalczyk (Eds.): TCCI XXII, LNCS 9655, pp. 1–9, 2016.
DOI: 10.1007/978-3-662-49619-0_1

Graphical scales "one to four" or "one to five" are often represented by stars, especially when used on the Internet, for rating movies, services, etc. In colloquial English, the idiomatic adjective "five-star" (meaning "first-rate") is frequently used. The choice of the rating scale upper limit as 5 may have something do with the number of fingers in one hand. The use of $m = 10$ may be influenced by the numerical system with 10 as the numerical base, which in turn is derived from the number of fingers on two hands. Rating scales can also include scores in between integers (or use graphics, such as a line to mark the answer with a vertical bar, ×, or another symbol) to give a more precise rating. The origin of rating scales are not clear but [1] seems to be one of the most cited (by, for example, the Web of Science count) and celebrated interpretations.

Pairwise comparisons have great application to collective intelligence since this method allows us to synthesize often highly subjective assessments of expert panels, steering committees, or other collective decision making constituencies. In case of doubt, it was evidenced in one of the flagship ACM publications [8]. Furthermore, Professor Kenneth J. Arrow, the Nobel prize winner, has used "pair" 24 times in his seminal work [3].

Finally, this study does not invalidate two major contributions in [7] regarding the scale selection and [5] regarding the rating scale unit used for pairwise comparisons. In fact, both studies and their contributors have a considerable impact on this study.

2   The Rating Scale Paradox

Paradoxes serve a very important purpose in science. They stimulate creative thinking. Banach Tarski paradox is one of the most stunning in mathematics. Russell's paradox contributed to a drastic paradigm shift in the foundation of set theory. This paradox calls for data entry correction. In the current situation, the relative error for the scale 1 to 9 (in Sect. 4) is 23 % for the value 2 which may be frequently entered since such a scale (with its own drawback) promotes the use of low and high values.

According to [2]:

> Graded responses are used where there is no measuring instrument of the kind found in the physical sciences, but the structure of the responses mirrors physical measurement. In physical measurement, the amount of a property of an entity is measured by using an instrument to map the property onto a continuum which has been divided into units of equal length, and then the count o f the number o f units from the origin that the property covers (often termed simply the measurement of the entity), is the location of the entity with respect to that property. Although it is recognized that instruments have operating ranges, the measurements of any entity are not taken to be a function of the operating range of an instrument–if the property exceeds the range of one instrument, then an instrument with a relevant range is sought. In deterministic theories, the variations and sizes of the property measured are considered sufficiently

large relative to the size of the unit, that errors of measurement are ignored, and the count is taken immediately as the measurement.

Using rating scales for the data entry is a popular method in most pairwise comparisons methods to collect graded responses or assessments. However, they suffer from the paradox and the acquired data cannot be entered into a PC matrix without a prior processing.

We need to clarify the terminology. Ratios of entities (sometimes they are referred to as "ratio scale values") create a PC matrix. The ratio scale is not the same the rating scale. Various rating scales are used for acquiring ratios but not all ratios are taken from a ratio scale.

We assume that M is a reciprocal PC matrix over $\mathbb{R}^+$. Let M be of the form:

$$M = \begin{bmatrix} 1 & m_{1,2} & \cdots & m_{1,n} \\ \frac{1}{m_{1,2}} & 1 & \cdots & m_{2,n} \\ \vdots & \vdots & \vdots & \vdots \\ \frac{1}{m_{1,n}} & \frac{1}{m_{2,n}} & \cdots & 1 \end{bmatrix}$$

The following simple example of a PC matrix for three entities A, B, and C:

$$M = \begin{bmatrix} 1 & 2 & 2 \\ 0.5 & 1 & 1 \\ 0.5 & 1 & 1 \end{bmatrix}$$

reflects $A = 2 * C$, $A = 2 * B$, and $B = C$ hence $A = 2, B = 1, C = 1$ is (one of many) solutions. So far, there seems to be no problem with this PC matrix since the above PC matrix M is consistent as the only triad in M fulfills the consistency condition:

$$m_{ij} \cdot m_{jk} = m_{ik} \tag{1}$$

for every $i, j, k = 1, 2, \ldots, n$.

There is one thing drastically missing in M: the rating scale upper limit. In other words, nothing will change if we use a different rating scale, say 1 to 3 (recommended in [10]). If so, we may try a bit bigger rating scale upper limit: 1 to 101 (giving 100 slots). For such a scale, things get a bit complicated as such a rating scale makes $A = B = C$ in practical terms, since 2 comes so close to 1 that it is hard to see it "with the naked eye" as we can imagine increasing the rating scale upper limit to infinity. Section 4 shows that even a scale, of a moderate size 1 to 9, causes a substantial approximation error. Regardless of the practicality, the rating scale values 1 and 2 become practically indistinguishable for large m. Let us recall that 1 on the rating scale stands for equality of compared entities. Using 2 for "two times bigger" (or somehow "superior") on even a moderate rating scale "1 to 10" has never been considered as incorrect yet weights (computed by any method since M is consistent) are:

$$[0.5, 0.25, 0.25]$$

It reflects the fact $A = 2*B = 2*C$, although $A = B = C$ with as high accuracy as we can wish to have for $m \to \infty$.
Evidently:

$$A = 2 * B$$
$$A = B$$

give $A * 1 = A * 2$ hence $1 = 2$ for $A > 0$ with as high accuracy as we wish to have.

This is a pairwise comparisons rating scale paradox of a fundamental importance since we provided evidence that: $A = 2 * B$ and $A = B$ for $A > 0$ and $B > 0$.

The paradox takes place since the entries in the PC matrix do not have any connection to the rating scale upper limit. The value 2 on the rating scale "1 to 10" is not the same 2 as on the scale 1 to 101. Evidently, the middle rating scale value depends on m and for $m = 101$, it is not 2 but 50.

3 Solution to the Paradox

Let us look at Fig. 1, representing value 2 on rating scales with the different upper limits. For $m = 101$, it is not "half of the rating scale". For $m = 9$, "the half" is 4. With the increased rating scale upper limit m, the meaning of the value 2 shifts towards "equality" with the relative error diminishing to 0. For $v \to \infty$ on a rating scale with $m \to \infty$, the value to be entered into the PC matrix is 2 since 1 is the "neutral" point.

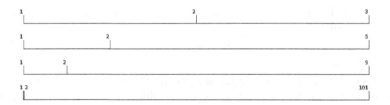

Fig. 1. Value 2 on various rating scales

We need to incorporate m into the PC matrix M. A normalizing mapping is proposed as a solution. Technically, the term *normalization* should not be used since we are not mapping the rating scale values to the interval $[0, 1]$ (open or close). In a PC matrix, reflecting the equality of all entities is done by all entries having a value of 1. Evidently, 0 (as a ratio) does not exist. The normalization of the ratio scale values into $[0, 1]$ interval and adding the "neutral" 1 prevents the paradox from taking place. For it, a linear function

$$f : [1, m] \to [1, 2]$$

such that $f(1) = 1, f(m) = 2$ needs to be defined (improved by [16]). It is given by:

$$f(x) = \frac{1}{m-1}x + \frac{m-2}{m-1}$$

Evidently, for $v \in [1, m]$, $f(v) = \frac{v+m-2}{m-1}$ hence $f(v) = 1 + \frac{v-1}{m-1}$. For a given value v on a scale "1 to m", the PC matrix entry should be:

$$1 + (v-1)/(m-1) \tag{2}$$

It is easy to see that:

$$\lim_{m \to \infty} 1 + (v-1)/(m-1) = 1$$

It means that our paradox no longer takes place. Certainly, other solutions may be considered and the future research is expected to contribute to it. The research of new "normalization" methods is in progress.

Example: The scale 1 to 6 is used in elementary schools in at least one of the EU countries. It has an interpretation for 2 as a *marginally passing mark*. It has a definitely different meaning than 2 on the scale 1 to 101 as a hypothetical scale for evaluating University students. In fact, assuming that 1 is the lowest score and 101 is the top score, it is hard to envision any school in any country setting 2 as satisfactory score.

Certainly, the rating scale upper limit of 101 can be extended to any arbitrarily large value bringing 2 as close to 1 as we can imagine. It does not matter whether or not we use such a scale since a substantial error occurs (23%) for even a relatively modest scale of 1 to 9 as evidenced by a numerical example in Sect. 4. ∎

It is also worth noticing that our normalizing mapping transforms rating scale values into $[1, 2]$ (and their inverses $[1/2, 1]$). It also creates PC matrices which have mathematically "nice" entries since their values are less than the Fülöp constant (approx. 3:330191) exploited in [10] to analyze the rating scale. The exact value of Fülöp's constant is equals to:

$$a_0 = ((123 + 55\sqrt{5})/2)^{1/4} = \sqrt{\frac{1}{2}\left(11 + 5\sqrt{5}\right)} \approx 3.330191 \tag{3}$$

Thanks to Fülöp's constant, the optimization problem for finding weights can be transformed into the convex programming problem with a strictly convex objective function to be minimized (see [9], Proposition 2):

$$\min \sum_{i=1}^{n-1} f_{a_{in}}(x_i) + \sum_{i=1}^{n-2}\sum_{j=i+1}^{n-1} f_{a_{ij}}(x_{ij})$$
$$\text{s.t.} \quad x_i - x_j - x_{ij} = 0, \ i = 1, \ldots, n-2, \ j = i+1, \ldots, n-1. \tag{4}$$

where the univariate function is defined as:

$$f_a(x) = (e^x - a)^2 + (e^{-x} - 1/a)^2 \tag{5}$$

and a is replaced by the Fülöp's constant (Fig. 2).

4 A Numerical Example

For a rating scale 1 to 9 (introduced to pairwise comparisons by Saaty in 1977 (for details, see [13])), value 2 gives what Fig. 3 demonstrates. The original and corrected values for this scale are in Table 1.

Table 1. The original and corrected values for the scale 1 to 9

1	2	3	4	5	6	7	8	9
1	1.125	1.25	1.375	1.5	1.625	1.75	1.875	2

Two 3×3 PC matrices with the original input data and the adjusted data by the geometric means (unnormalized and normalized to 1) are illustrated by Fig. 3. Bars on the left demonstrate the original data. Bars on the right are for the normalized data.

1	2	2
0.5	1	1
0.5	1	1

1.59	0.5
0.79	0.25
0.79	0.25

1	1.25	1.25
0.8	1	1
0.8	1	1

1.16	0.37
0.93	0.29
0.93	0.29

Fig. 2. PC matrices with the original input data and the adjusted data

The relative error for the above results:

$$\eta = \frac{\epsilon}{|v|} = \left| \frac{v - v_{\text{approx}}}{v} \right| = \left| 1 - \frac{v_{\text{approx}}}{v} \right|,$$

is: 23 % for all three entities which can hardly be ignored. Pretending that "nothing happened" is not an option when such a scale is used for a project of national importance (e.g., safety of a nuclear power station).

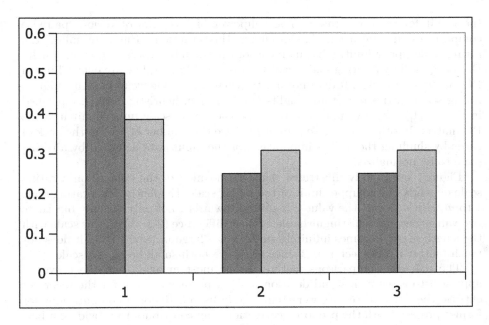

Fig. 3. Weights for the original and corrected data

5 Input Data Without the Paradox Effect

Evidently, rating scale entries cannot be directly entered into a PC matrix without prior processing. The proposed normalization prevents the paradox from taking place. In particular, experiments with randomly generated bars in [11, 12] render not only the correct results but evidence that the estimation error decreases when pairwise comparisons are used.

The Stone Age, mentioned in this author's former publications, and the ratio of stone weights give also the correct entries for the direct inclusion into a PC matrix since no rating scale is involved in it. The same goes for other physical measurements (distance, time) which include 0. However, there is a problem with the temperature expressed in Celsius scale as 0 is not the absolute 0 as in Kelvin (correct) scale.

It is also important to point out that most fuzzy extensions of PCs (of which the most cited in Web of Science is [15]) do not suffer from the presented paradox since they have a membership function with values in [0, 1]. However, it also requires further scientific examination.

6 Conclusions

Entering categorical data into a PC matrix, without prior preprocessing, leads to a paradox. The absence of the rating scale upper limit in a PC matrix is the result for such a paradox. The willpower of inventors of various pairwise comparisons customizations cannot change this situation. "Having in mind" the rating scale upper limit, without incorporating it into processing, cannot help. Data acquired by a rating scale must be processed before they are entered into a PC matrix for weights. It does not matter what processing we use for the original rating scale entries to obtain weights (for example, heuristic methods specified in [17] or [4]) when we fail to incorporate the rating scale upper limit into the PC matrix. In other words, we cannot increase the number of aces in the deck of cards by shuffling them. It is imperative for the input data acquired by a rating scale to be normalized.

Figure 1 adequately illustrates that the meaning of the value 2 on a rating scale depends on the upper limit of the rating scale. Evidently, for a rating scale 1 to m with $m \to \infty$, the value 2 is shifted towards 1 not only visually but these two values become indistinguishable as their difference (2-1=1) on this scale with the length $m - 1$ becomes infinitely small. For a large m (say, 10^{999}), it does not much matter if we select 1 or 2 since these are both small on such a scale.

The pairwise comparisons method is the most amazing and the universal approach to assessments and decision making problems. Even for the incorrect entries, the received results were remarkably useful. Redoing computations for former projects with the proposed correction may contribute to providing a better evidence of the superiority of the pairwise comparisons method in terms of higher precision. The bad news is that most former publications on pairwise comparisons should be redone if they used data acquired by a rating scale unless they preprocessed input data by a method which prevents the paradox from occurring. The good news for authors of such studies and University administrators is that they may improve their publication record.

Acknowledgments. This project has been supported in part by the Euro Research grant "Human Capital". The author is grateful to Grant O. Duncan (Team Lead, Business Intelligence and Software Integration, Health Sciences North, Sudbury, Ontario) for his help with proofreading this text. Matteo Brunelli and William C. Wedley (listed in the alphabetical order) have commented on the first draft and inspired many enhancements of this presentation. For it, the author is very grateful to them. The author also acknowledges that improvements and extensions to the Author's original mathematical formulas ([16]), done by Eliza Wajch, are of great importance. Numerous researchers on four continents (Australia, Asia, Europe, and North America) have been extremely supportive through this project and I would like to thank all of them.

I owe thanks most of all to Prof. Kenneth Arrow, the Nobel Laureate, for reading this text (posted on ResearchGate.net).

References

1. Andrich, D.: A rating formulation for ordered response categories. Psychometrika **43**, 357–374 (1978)
2. Andrtch, D.: Models for measurement, precision, and the nondichotomization of graded responses. Psychometrika **60**(1), 7–26 (1995)

3. Arrow, K.J.: A difficulty in the concept of social welfare. J. Polit. Econ. **58**(4), 328–334 (1950)
4. Kulakowski, K.: A heuristic rating estimation algorithm for the pairwise comparisons method. Cent. Eur. J. Oper. Res. **23**(1), 187–203 (2015)
5. Choo, E.U., Wedley, W.C.: Estimating ratio scale values when units are unspecified. Comput. Ind. Eng. **59**(2), 200–208 (2010)
6. Davis, H.A.: The Method of Pairwise Comparisons. Griffin, London (1963)
7. Dong, Y.C., Xu, Y.F., Li, H.Y., Dai, M.: A comparative study of the numerical scales and the prioritization methods in AHP. Eur. J. Oper. Res. **186**(1), 229–242 (2008)
8. Faliszewski, P., Hemaspaandra, E., Hemaspaandra, L.A.: Using complexity to protect elections. Commun. ACM **53**(11), 74–82 (2010)
9. Fülöp, J.: A method for approximating pairwise comparison matrices by consistent matrices. J. Global Optim. **42**, 423–442 (2008)
10. Fülöp, J., Koczkodaj, W.W., Szarek, S.J.: A different perspective on a scale for pairwise comparisons. In: Nguyen, N.T., Kowalczyk, R. (eds.) Transactions on Computational Collective Intelligence I. LNCS, vol. 6220, pp. 71–84. Springer, Heidelberg (2010)
11. Koczkodaj, W.W.: Statistically accurate evidence of improved error rate by pairwise comparisons. Percept. Mot. Skills **82**(1), 43–48 (1996)
12. Koczkodaj, W.W.: Testing the accuracy enhancement of pairwise comparisons by a Monte Carlo experiment. J. Stat. Plann. Infer. **69**(1), 21–31 (1998)
13. Koczkodaj, W.W., Mikhailov, L., Redlarski, G., Soltys, M., Szybowski, J., Tamazian, G., Wajch, E., Yuen, K.K.F.: Important facts and observations about pairwise comparisons. Fundamenta Informaticae **144**, 1–17 (2016)
14. Thurstone, L.L.: A law of comparative judgements. Psychol. Rev. **34**, 273–286 (1927)
15. Vanlaarhoven, P.J.M., Pedrycz, W.: A fuzzy extension of saatys priority theory. Fuzzy Sets Syst. **11**(3), 229–241 (1983)
16. Wajch, E.: Private Communication (by email), 15–27 November 2015
17. Williams, C., Crawford, G.: Analysis of subjective judgment matrices, The Rand Corporation Report R-2572-AF, pp. 1–59 (1980)

On Achieving History-Based Move Ordering in Adversarial Board Games Using Adaptive Data Structures

Spencer Polk$^{(\boxtimes)}$ and B. John Oommen

School of Computer Science, Carleton University, Ottawa K1S 5B6, Canada
andrewpolk@cmail.carleton.ca, oommen@scs.carleton.ca

Abstract. This paper concerns the problem of enhancing the well-known alpha-beta search technique for intelligent game playing. It is a well-established principle that the alpha-beta technique benefits greatly, that is to say, achieves more efficient tree pruning, if the moves to be examined are ordered properly. This refers to placing the best moves in such a way that they are searched first. However, if the superior moves were known *a priori*, there would be no need to search at all. Many move ordering heuristics, such as the Killer Moves technique and the History Heuristic, have been developed in an attempt to address this problem. Formerly unrelated to game playing, the field of Adaptive Data Structures (ADSs) is concerned with the optimization of queries over time within a data structure, and provides techniques to achieve this through dynamic reordering of its internal elements, in response to queries. In earlier works, we had proposed the Threat-ADS heuristic for multi-player games, based on the concept of employing efficient ranking mechanisms provided by ADSs in the context of game playing. Based on its previous success, in this work we propose the concept of using an ADS to order moves themselves, rather than opponents. We call this new technique the History-ADS heuristic. We examine the History-ADS heuristic in both two-player and multi-player environments, and investigate its possible refinements. These involve providing a bound on the size of the ADS, based on the hypothesis that it can retain most of its benefits with a smaller list, and examining the possibility of using a different ADS for each level of the tree. We demonstrate conclusively that the History-ADS heuristic can produce drastic improvements in tree pruning in both two-player and multi-player games, and the majority of its benefits remain even when it is limited to a very small list.

Keywords: Alpha-beta search · Adaptive data structures · Move ordering · History heuristic · Killer moves

The second author is grateful for the partial support provided by NSERC, the Natural Sciences and Engineering Research Council of Canada. A preliminary version of this paper was presented at ICCCI'15, the *7th International Conference on Computational Collective Intelligence Technologies and Applications*, in Madrid, Spain, in September 2015.

N.T. Nguyen and R. Kowalczyk (Eds.): TCCI XXII, LNCS 9655, pp. 10–44, 2016.
DOI: 10.1007/978-3-662-49619-0_2

1 Introduction

The problem of achieving robust game play, in a strategic board game such as Chess or Go, against an intelligent opponent is a canonical one within the field of Artificial Intelligence (AI), and has seen a great deal of research throughout the history of the field [20,25]. A substantial portion of the vast body of literature present in this field is based on the highly effective alpha-beta search technique, which provides an efficient way to intelligently search a potentially very large number of moves ahead in a game tree, while pruning large sections of the tree which have been found to be irrelevant [7,20]. Using this technique, great strides have been made over the years in competitively playing many strategic board games at the level of top human players.

It is well known that the performance of the alpha-beta search is greatly impacted by a proper move ordering. This involves arranging possible moves so that the best move is likely to be searched first. Based on this knowledge, a substantial body of literature exists that spans a wide variety of move ordering heuristics that attempt to achieve this [7,17,23]. Examples of these techniques include the well-known Killer Moves strategy, and the History Heuristic, which serve as domain-independent approaches, that operate by remembering those moves that have performed well earlier in the search, and prioritizing them later [23].

The formerly unrelated field of Adaptive Data Structures (ADSs) is concerned with the problem of query optimization within a data structure, based on the knowledge that not all elements are accessed with the same frequency [3,5,6]. This problem is addressed through dynamic reorganization of the data structure's internal order, in an attempt to place elements accessed with a higher frequency nearer to the head of the list [1]. This reordering is accomplished in response to queries as they are received, and the field of ADSs proposes a number of possible mechanisms by which a data structure can be reordered in response to the queries, such as the Move-to-Front or Transposition rules for adaptive lists [1,5,6].

Observing that there is an intuitive link between the dynamic reordering of elements of an ADS in response to queries, and move ordering strategies in games, we had previously proposed the Threat-ADS heuristic, for multi-player games, which employs an adaptive list to rank *opponents* based on their relative threats [12]. The specific case of multi-player game playing is relatively unexplored in the literature dealing with intelligent game playing, and poses a number of unique challenges, which prevent existing multi-player techniques from achieving a performance comparable to their two-player counterparts [9,22,28–30]. The Threat-ADS was shown to be able to achieve statistically significant gains in terms of tree pruning, in a wide range of configurations, by considering different ADS update mechanisms and starting positions of the game [13,15].

Based on the success of the Threat-ADS heuristic in the multi-player domain, we hypothesize that ADS-based ranking may be applied in other areas in the context of game playing. Specifically, based on the Killer Moves and History Heuristic techniques, we propose a related move ordering heuristic, which we refer to as the History-ADS heuristic, which uses the qualities of an ADS to rank individual moves, augmenting their position in the ADS when the move is found to produce a cut. In this work, we show that, while using lightweight, efficient ranking techniques associated with an ADS, the History-ADS is able to obtain substantial gains in tree pruning in both the two-player and multi-player cases, in a variety of games.

Preliminary results related to this work were presented in [14, 16]. The remainder of the paper is laid out as follows. Section 2 presents a background on game playing, and a fairly brief description of the Killer Moves and History Heuristic techniques. Section 3 introduces the field of ADSs, and the techniques from that field that we employ in this work. Section 4 details our previous work, the Threat-ADS heuristic, based on which we describe the novel History-ADS heuristic. Section 6 describes possible refinements to the History-ADS. Section 7 describes our experimental configuration and game models, and Sects. 8, 9, 10 and 11 present our results. Section 12 provides our discussion and analysis of these results, and Sect. 13 concludes the paper.

2 Game Playing Background

Historically, the primary technique for achieving competent game play against an adversarial opponent has been based on the Minimax strategy, which has been successfully applied to the problem of intelligent game playing from the theoretical roots of the discipline, to the modern era [10, 20, 25]. The Minimax strategy, as its name implies, attempts to maximize the perspective player's possible gain, when considering each possible move or action he could take, while assuming that the opponent does the opposite, or attempts to minimize the perspective player's returns.

When applied to a two-player, turn-based board game, such as Chess, the Minimax technique achieves intelligent and informed game play by searching a number of moves ahead in the game tree that represents all possible paths of moves in the game, or to a given depth, usually referred to as a ply [20]. The game tree is explored in a depth-first manner, until the desired ply is reached, at which point the game state is evaluated according to some form of refined heuristic, assigning a value to the position for the max, or perspective, player [20]. These values are then passed upwards through the tree, assuming that in positions where the perspective player is making a move, the maximum of these values will be selected, and the minimum will be chosen when the opponent can make a decision. In normal games, these options generally alternate with each level of the tree. Upon completion of the search, the root of the tree, representing the current turn, is assigned a value. This represents our best estimation, according to our available search depth and heuristic, of the best possible move available to the perspective player. A simple example of a game tree explored according to the Minimax strategy is presented in Fig. 1.

From the above explanation, one can intuitively see that the strength of the Minimax technique is dependent upon two major factors. The first of these

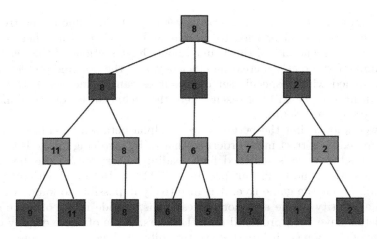

Fig. 1. A simple example of a game tree explored according to the Minimax strategy. The red nodes represent the perspective or max player, and blue nodes represent the opponent, or min player.

is the strength of the evaluation heuristic employed in leaf positions, as a weak heuristic will necessarily lead to an ill-informed understanding of the game state, while the use of highly refined heuristics employing expert knowledge is a standard method of insuring the strong performance of a game engine [10,20,22]. The other factor is the maximum possible ply depth that the engine can search to, since the ability to search deeper represents a higher degree of lookahead in the game. This allows for the formulation of more complex strategies, and takes care of avoiding possible pitfalls or traps set by the opponent. Maximum achievable search can be increased, as expected, through improvements to available hardware, or refinements to the Minimax technique.

Over the course of its history, a number of refinements, modifications, and improvements to the Minimax algorithm have been proposed, including techniques to improve gameplay logic, such as quiescence search, and extensions of the Minimax technique to environments other than those that involve perfect information for two player strategic games, such as multi-player environments, and games of incomplete information [20,28]. However, a major focal point of improvements to the Minimax technique is in achieving a greater lookahead, via a more efficient search, including arguably the most well-known enhancement, alpha-beta pruning.

The well-known alpha-beta search (which refers to a Minimax search employing alpha-beta pruning) is based upon the observation that not all moves available at different levels of the game tree will impact the its value. Some of them are so poor that they will never be reasonably selected, while others are so strong that the opponent will never allow a situation where they can come to pass [7]. Furthermore, it is possible, through the construction of upper and lower bounds on the possible values at a given node, commonly called the alpha and beta values, to prove that a given node can never impact the value of the tree, and thus, its children no longer need to be searched [7]. Given that there may be many possible descendants of this pruned node, huge sections of the search tree can be eliminated, greatly increasing efficiency, and the possible ply depth that can be searched with a specific set of resources can be increased. Due to its well-known nature, we will not discuss here the technicalities of the alpha-beta search in any greater detail.

It is well-known that the performance of alpha-beta search can be substantially improved by correct move ordering, that is, by involving methods by which the best possible move is searched first, leading to stricter bounds being constructed, and thus, more efficient pruning [20,23,28]. However, without perfect information about the game tree, it is intuitively impossible to know, with certainty, the identity of the superior moves. Thus, a wide range of move ordering heuristics have been proposed over the years. Some of these employ expert knowledge of the game to insure that strategically good moves are examined first. Others are domain independent strategies that apply to a wider range of strategic board games [10,22–24]. Two well-known examples of these, upon which the present work is based, are described in detail in the following section.

2.1   The History Heuristic and Killer Moves

The number of techniques available to achieve efficient move ordering in a game playing engine is exhaustive, and to fully detail every one available in the literature would be outside the scope of this work. In this work, we specifically consider two well-known, commonly-used move ordering heuristics, which are the Killer Moves heuristic, and the History heuristic [23]. These techniques are related, in that both attempt to remember effective moves encountered ("effective" being defined as those likely to produce a cut, resulting in a smaller tree), and to explore them first if they are encountered elsewhere in the tree. Indeed, the History heuristic is regarded as a generalization of the Killer Moves heuristic, from a local to a global environment, within the tree [23]. However, both are still commonly employed in modern game engines [10,22,23].

The Killer Moves heuristic (also sometimes called the Killer heuristic) operates by prioritizing moves that were found to be good (that is, that produced a cut) in sibling nodes. For example, in the case of Chess, if it was found at some level of the game tree that White moving a bishop from C1 to A3 produced a cut, and that same move is encountered in another branch at the same level of the tree, it will be examined before other moves [23]. The heuristic is based on the assumption that each move does not change the board state that much. Therefore, if a move produced a cut in another position, it is likely to do well elsewhere, even if the preceding moves are different. Of course, this means that the Killer Moves heuristic can potentially be less effective in games where single moves do in fact produce large changes within the game.

The Killer Moves heuristic accomplishes this prioritization by maintaining a table in memory that is indexed by the depth. Within each memory location, a small number of "killer" moves are maintained (usually two), in a linked list or similar data structure [23]. If a new move produces a cut at a level of the tree where the list is full, older moves are replaced according to some arbitrary replacement scheme. When new moves are encountered, the "killer" moves in the table at the current depth are analyzed first, if they are applicable. Note that, as the algorithm can check if the killer moves are available when expanding a new node, and examine them immediately, the Killer Moves heuristic does not require the nodes to first be sorted. In fact additional time can be saved by not even generating the remaining moves if one of the killer moves produces a cut.

The History Heuristic is an attempt to apply the Killer Moves heuristic on a global scale, allowing moves from other levels in the tree to influence decisions. While a simplistic approach would be to maintain only a single list of "killer" moves and to apply it at all levels of the tree, this would allow moves that produce cuts near the leaves (as there will be many more of them, due to the explosive nature of the game tree), to have a disproportionate effect on the moves within the list [23]. The history heuristic therefore employs a mechanism by which cuts produced higher in the game tree have a greater impact on deciding which move to analyze first.

This is accomplished by maintaining a large array, usually of three dimensions, where the first index is 0, for the maximizing player, and 1, for the minimizing player. The next two indices indicate a move in some way. For example, in the case of Chess, the array is normally indexed by [from][to] where each of [from] and [to] are one of the 64 squares on the board. Within each of these cells is a counter, which is incremented when the corresponding move is found to produce a cut [23]. This counter is incremented by the value $depth * depth$, or 2^{depth}, thereby insuring the value increases more if the cut is higher in the tree [23]. When moves are generated, they are ordered by their value in this array, from greatest to least. In this way, moves that have produced a cut more often, and moves that produced cuts higher in the tree, are examined first.

The History heuristic is noted as a particularly effective and efficient move ordering technique [23]. However, it does have some drawbacks. The array of counters it must store is relatively large, although not a practical concern for modern computers, being two 64-by-64 arrays in the case of Chess. More importantly, unlike the Killer Moves heuristic, moves cannot be generated from the History heuristic; they must be sorted, adding non-linear time at every node in the tree. Thus, it is desirable to look at other heuristics if they are capable of cutting branches of the tree without adding this sorting time. Furthermore, in very deep trees, the History heuristic is known to become less effective, to the point where some modern Chess-playing programs, in particular, either do not use it or limit its application [24].

3 Adaptive Data Structures

It is a well-known problem, in the field of data structures, that the access frequencies of elements within a data structure are not uniform [5,6]. As an illustrating example, consider a linked list consisting of five elements, A, B, C, D, and E, in that order, where the corresponding access probabilities are 20 %, 5 %, 10 %, 40 % and 25 %. Using a traditional singly-linked list, these access probabilities pose a problem, as the two elements accessed the most frequently, D and E, are located at the rear of the list, thus requiring a longer access time. We can intuitively see that another linked list, holding the same five elements, in the order D, E, A, C, B will achieve faster average performance. Thus, by restructuring the list, one can obtain an improved functionality for the data structure.

In the trivial example above, the reorganization is obvious, as the access probabilities are assumed to be known and stationary. However, in the real world, the access probabilities are, as one would expect, not known when the structure is first created. The field of ADSs concerns itself with finding good resolutions to this problem [1,3,5,6]. As the access probabilities are not known, the data structure must learn them as queries proceed, and *adapt* to this changing information by altering its internal structure to better serve future queries [5]. Individual types of ADSs provide different methods to achieve this sort of behaviour for the specific data structure. An ADS may be of any type, typically a list or tree, with the well-known Splay Tree being an example of the second type [1,11]. However, in this work, we will be focusing exclusively on adaptive lists.

The method by which an ADS reorganizes its internal structure, in response to queries over time, must logically possess several qualities in order to be useful. Specifically, as the goal of an ADS is to improve the amortized runtime, by allowing more frequently accessed elements to be queried faster, the mechanism by which it reorders itself must itself be very efficient, or time lost on its execution would render benefits to query time irrelevant. Thus, methods developed in the fields of ADSs are typically simple, constant-time operations that do not require many memory accesses, comparison, or the use of counters.

In our previous work, we observed that the specific qualities of ADSs enable an ADS to be used as a highly efficient, dynamic *ranking* mechanism for other domains in game playing, provided two requirements can be met. The first is that the objects that we wish to rank can be represented in some way by the elements of the data structure, where the internal structure of the ADS can be seen to reflect their relative ranking. The second is that some method needs to exist to query the ADS when one of the ranked elements should be moved closer to the top position.

Given the wide range of potential objects that can be ranked within the domain of game playing, especially in the context of move ordering, we previously proposed that techniques from ADSs could be applied as an improving agent in the formerly-unrelated domain of game playing. This innovation led to the development of the Threat-ADS heuristic for multi-player games, where an ADS was employed to rank opponents, and this information was used to achieve move ordering in a state-of-the-art, multi-player technique. The Threat-ADS heuristic is described in more detail below.

3.1   ADS Update Mechanisms

The field of ADSs provides a wide range of techniques by which an ADS can reorganize its internal structure. We shall refer to these as *update mechanisms*. As alluded to in the previous section, these techniques tend to be very efficient, with a few constant-time operations, generally consisting of swapping the locations of a small number of elements in the list. In our previous work, we examined a wide range of known, ergodic, ADS update mechanisms, and found that they performed roughly equally well when applied to move ordering [13]. We have thus, to avoid repetition, restricted our analysis in this work to two very well known and frequently contrasted ADS update mechanisms, i.e., the Move-to-Front and Transposition rules.

Move-to-Front: The Move-to-Front update rule is one of the oldest and most well-studied update mechanisms in the field of ADSs [1,5,19,26]. Not coincidentally, it is also one of the most intuitive. As its name suggests, when an element is accessed by a query, in a singly-linked list, it is moved to the head, or front, of the list. Thus, if an element is accessed with a very high frequency, it will tend to stay near the front of the list, and therefore will be less expensive to access. It is also intuitive to see that, for a list where elements have $O(1)$ pointers to the next element, performing the action of moving an element to the front of the list is also $O(1)$, thus making the update mechanism very inexpensive to implement.

Given that elements are always moved to the front of the list when using the Move-to-Front rule, the list changes quite dramatically in response to each query, and this can cause it to generate more expensive queries compared to its competitors in many circumstances [19]. However, unlike its competitors, the Move-to-Front update mechanism provides the valuable property of a lower bound on cost in relation to the optimal ordering. It has been shown that the Move-to-Front update mechanism will provide a system that costs no more than twice that of the optimal ordering [1,5]. This guarantee insures the Move-to-Front rule remains attractive, even when compared to competing update mechanisms, which can often outperform it.

Transposition: The most common competitor to the Move-to-Front rule, also studied extensively in the ADS literature, is the Transposition rule [1,4,5]. It is no more difficult to implement or understand than the Move-to-Front rule, and, like its chief competitor, offers a powerful performance gain with interesting properties. When an element is accessed, under the Transposition rule, it is *swapped* with the element immediately ahead of it in the list. Thus, as an element is accessed more and more frequently, it will slowly approach the head of the list, contrasted with Move-to-Front, where it is immediately placed there.

As can be deduced from its behaviour, the Transposition rule is less sensitive to change than the Move-to-Front rule, which, depending on the problem domain, can be a good or bad thing [1]. Under many circumstances, the Transposition rule will be much closer to the optimal rule than the Move-to-Front rule over a long period of time [4,19]. Unfortunately, the Transposition rule does not offer any lower bound on cost in relation to the optimal ordering, and arguments have been made for either it or the Move-to-Front rule in different domains, leading to a historical lack of consensus in the field [1,4,26]. It is thus natural, when exploring a new domain of applicability with ADSs, to examine these two contrasting rules.

4   Previous Work: The Threat-ADS Heuristic

Our previous work focused exclusively on the domain of Multi-Player Game Playing (MPGP), which is a variant on traditional two-player game playing, where the number of opponents is greater than unity. Although multi-player games can be thought of as a generalization of the two-player environment to an N-player case, the majority of research has continued to focus on two-player games, with a substantially lesser focus on MPGP being present in the literature [9,22,27,28,31]. However, through the addition of multiple self-interested agents, such games present a number of complications and challenges that are not present in traditional two-player game playing. These include:

- One player's gain does not necessarily generate an equal loss amongst all opponents.
- Temporary coalitions of players can arise, even in games with only a solitary winner.
- The board state can change more between each of the perspective player's moves.
- A single-valued heuristic is not always sufficient to correctly evaluate the game state.
- Established, highly-efficient tree pruning techniques, such as alpha-beta pruning, are not always applicable.

Despite these challenges, due to the historical success of Mini-Max with alpha-beta pruning, in a wide range of environments, the majority of MPGP strategies have been based on its extensions to a multi-player environment [9,22,27,28]. These include the Paranoid and Max-N algorithms, which operate by assuming a coalition of opponents against the perspective player, and by extending the heuristic to a tuple of values, one for each player, and where one assumes that each agent seeks to maximize his own value [28]. The details of these algorithms are omitted here in the interest of brevity and relevance.

In recent years, a novel MPGP technique, named the Best-Reply Search (BRS), has been proposed, which is capable of achieving substantially stronger performance than either the long-standing Paranoid or Max-N algorithms in a wide variety of environments [22]. Given its state-of-the-art nature, our previously proposed technique, the Threat-ADS heuristic, was designed with it in mind. The BRS, and the Threat-ADS heuristic, are detailed in following sections, as the Threat-ADS forms the basis of the new work that we present here.

4.1   Best-Reply Search

The BRS attempts to simplify the problem of a multi-player game back to a two-player game, to take advantage of the breadth of techniques available in the two-player context, and avoid the extremely large search space the Paranoid and Max-N algorithms must consider [22]. It achieves this by grouping all opponents together, and considering them to be a single, "super-opponent". During each Min phase of the tree, this "super-opponent" is only allowed to make a single move, or, in other words, only one opponent is permitted to act. This opponent is the one who has the most minimizing move, in relation to the perspective player, at this point in time, or the "Best Reply". Figure 2 shows a single level of a BRS tree (only a single level is shown for space considerations, as the branching factor is considerably higher for opponent turns in BRS), where the minimum of all opponent turns is being selected.

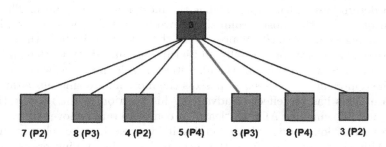

7 (P2)   8 (P3)   4 (P2)   5 (P4)   3 (P3)   8 (P4)   3 (P2)

Fig. 2. The operation of a single level of the Best-Reply Search. The scores that are reported have the opponent's player number listed next to them (in parenthesis) to assist in the clarification.

The immediate, glaring drawback of the BRS algorithm is that it considers illegal move states while searching. This is certainly a serious drawback, and in fact, limits the games to which BRS can be applied [22]. BRS can only be applied to those games where it is *meaningful* for players to act out of turn, and performs best when the board state does not change too dramatically in between turns [22]. An example of a game to which BRS can *not* by applied is Bridge, because scoring in Bridge is based on tricks, and thus, allowing players to act out of order renders the game tree to be void of meaning. In a game where the game state changes significantly between turns, there is a serious risk of the BRS arriving at a model of the game which is significantly different from reality.

However, in cases where it can be applied, the BRS has many benefits over the Paranoid and Max-N algorithms, and often outperforms them quite dramatically [22]. By considering the multi-player game as if there were two players, issues related to pruning and potential lookahead for the perspective player are mitigated, which can lead to much better game play in certain games where the game state does not change much during each turn, such as Chinese Checkers, but where many opponents may be present [22].

4.2   The Threat-ADS Heuristic

A factor that is present in multi-player games, but which has no equivalent in the two-player case, is that of relative opponent threat, which we define as a ranking of opponents based on their potential to minimize the perspective player, either based on their current board position, or some knowledge we have gained about their skill. The idea of of considering opponent threat, both to model and predict their future actions, as well as to prioritize opponents, is a known concept within extant multi-player game playing literature [29,30,32]. As the BRS groups all opponents together, and must consider all possible moves they could make to find the best one, it presents an opportunity where a ranking of opponents based on their relative threats can be applied to move ordering. This is precisely how the Threat-ADS functions, employing an ADS to enable this ranking to take place.

Considering the execution of the BRS, we observe that, at each "Min" phase, the BRS determines which opponent has the most minimizing move. The phenomenon of having the most minimizing move against the perspective player, and also being characterized by possessing the relatively highest threat level against that player, are conceptually and intuitively linked. With that in mind, we query an ADS, which contains the identities of each opponent, with the identity of the opponent that is seen to possess the most minimizing move during a Min phase. This has the effect of advancing his position in the relative threat ranking, and allowing the ADS to "learn" a complete ranking over time.

With this ranking provided, we employ it by exploring the relevant moves, at each Min phase, in order from the most to least threatening opponent. As a threatening opponent is more likely to provide the greatest minimization to the perspective player, this improves move ordering, and thus the savings from alpha-beta pruning. We clarify this by means of an example in Fig. 3. This figure shows how the ADS updates, based on which opponent was found to have the most minimizing move, at a certain level of the tree. Here, opponent "P4" has the most minimizing move, and thus the ADS is updated by moving him to the head of the list.

We observe that the Threat-ADS heuristic is particularly lightweight, requiring only efficient, constant-time updates, and retains a list of a size equal to the number of opponents, which is likely to be a very small constant (typically no more then seven). Furthermore, the Threat-ADS heuristic has the quality of not requiring the moves to be sorted, similar to Killer Moves, as they may simply be generated in the order of the ADS. In our previous work, we demonstrated that the Threat-ADS heuristic was capable of producing meaningful, statistically significant gains in terms of tree pruning in a variety of multi-player games, and employing a range of update mechanisms, at different points in time within the game's progression, and for a variety of opponents [12,13,15].

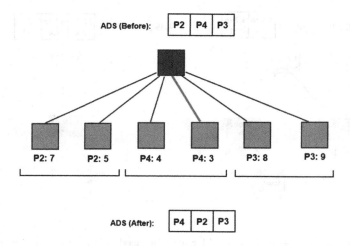

Fig. 3. A demonstration of how the Threat-ADS heuristic operates over time.

5   The History-ADS Technique

As mentioned earlier, the novel technique we propose in this work is inspired by our previous Threat-ADS heuristic, and by the well-known Killer Moves and History Heuristic move ordering strategies. Specifically, we wish to employ the same metric for achieving move ordering that the established techniques employ, that of move history. Thus, we intend to create a ranking of moves based upon their previous performance within the search, or more specifically, by providing a higher ranking to those that have produced cuts previously. We then prioritize those moves which possess a higher rank when they are encountered later in the search, with the expectation that those that have produced a cut before will be more likely to do so again, leading to improvements in the efficiency of the alpha-beta search technique. As with the Threat-ADS heuristic, we wish to accomplish this ranking by means of an ADS, given that ADSs provide efficient and dynamic ranking mechanisms.

Rather than utilizing an ADS whose elements are opponents, as in the case of the Threat-ADS, we, instead, employ an ADS containing moves. However, unlike the case with the Threat-ADS, we begin with an empty adaptive list. When a move is found to produce a cut, we query the ADS with the identity of that move. To illustrate this, in the case of Chess, we would query it with the co-ordinates of the square that the piece originated from, and the co-ordinates of its destination, as one does in invoking the History heuristic. If the ADS already contains the move's identity, its position within the ADS is changed according to the ADSs' update mechanisms. If it is *not* within the ADS, it is instead appended to the end, and immediately moved as if it were queried.

An example of how the ADS can manage move history over the course of the game is depicted in Fig. 4, which showcases its learning process and application over a fragment of the search.

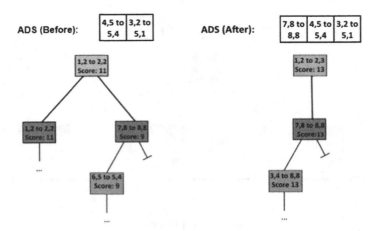

Fig. 4. A demonstration of how an ADS can be used to manage move history over time. The move (7,8) to (8,8) produces a cut, and so it is moved to the head of the list, and informs the search later.

5.1   Specification of the History-ADS Heuristic

The precise execution of the History-ADS heuristic is very similar to that of our previously-introduced Threat-ADS heuristic. First of all, we must understand how to update the ADS at an appropriate time, i.e., through querying it with the identity of a move that has performed a cut. This is analogous to querying the ADS with the most threatening opponent within the context of the Threat-ADS. Then, at each Max and Min node, we must somehow order the moves based on the order of the ADS.

Fortunately, within the context of alpha-beta search, there is a very intuitive location to query the ADS, which is where an alpha or beta cutoff occurs, before terminating that branch of the search. This is, of course, analogous to the timing with which the History Heuristic updates its structure. To actually accomplish the move ordering, when we expand a node and gather the available moves, we explore them in the order proposed by the ADS, again, similarly to how moves are ordered by their value according to the History Heuristic.

The last issue that must be considered is that, unlike in the Threat-ADS where opponent threats were only relevant on Max nodes, when considering move history, the information is relevant on both Max and Min nodes. Furthermore, we observe that a move that produces a cut on a Max node may not be likely to produce a cut on a Min node, and vice versa. This would occur, for example, if the perspective player and the opponent do not have analogous moves, such as in Chess or Checkers, or if they are some distance from each other. We thus employ two list-based ADSs within the History-ADS heuristic, one of which is used on Max nodes, while the other is used on Min nodes. Algorithm 1 shows the Mini-Max algorithm, employing the History-ADS heuristic.

Algorithm 1. Mini-Max with History-ADS

Function BRS(node, depth, player)

1: if node is terminal or depth ≤ 0 then
2:     **return** heuristic value of node
3: else
4:     if node is max then
5:         for all child of node in order of **MaxADS** do
6:             $\alpha = max(\alpha, minimax(child, depth - 1)$
7:             if $\beta \leq \alpha$ then
8:                 break (Beta cutoff)
9:                 query MaxADS with cutoff move
10:            end if
11:        end for
12:        return α
13:    else
14:        for all child of node in order of **MinADS** do
15:            $\beta = min(\alpha, minimax(child, depth - 1)$
16:            if $\beta \leq \alpha$ then
17:                break (Alpha cutoff)
18:                query MinADS with cutoff move
19:            end if
20:        end for
21:        return β
22:    end if
23: end if

End Function Mini-Max with History-ADS

5.2 Qualities of the History-ADS Heuristic

As emphasized earlier, the History-ADS heuristic is very similar in terms of construction to the Threat-ADS heuristic. As with the Threat-ADS, it does not in any way alter the final value of the tree, and thus cannot deteriorate the decision-making capabilities of the Mini-Max or BRS algorithms. Similar to the Threat-ADS, the ADSs are added to the search algorithm's memory footprint, and their update mechanisms with regard to its running time.

Compared to the Threat-ADS, the History-ADS can be expected to employ a much larger data structure, as there will be many more possible moves than total opponents in any non-trivial game. Furthermore, as illustrated above, we maintain two separate ADSs in the case of the History-ADS, which rank minimizing and maximizing moves, respectively. The History-ADS, as described here, thus remembers any move that produces a cut within its ADS for the entire search, even if it never produces a cut again and lingers near the end of the list. Further, we emphasize that new moves could be regularly added to the corresponding lists. However, while this may appear to suggest that the data structures are of unbound size, depending on how we identify a move, there are, in fact, a limited number of moves that can be made within a game.

Revisiting the example of Chess, if, as in the case of the History heuristic, we consider a move to be identified by the co-ordinates of the square in which the moved piece originated, and the co-ordinates of its destination square, we see that there are a maximum of 4096 possible moves. This figure serves as an upper bound on the size of the ADS. In the case of the History heuristic, an array of 4096 values, or whichever number is appropriate for the game, is maintained for both minimizing and maximizing moves, whereas the History-ADS only maintains information on those moves that have produced a cut. Unarguably, this is a significantly smaller subset, in most cases. We thus conclude that the memory requirements of the History-ADS are upper bounded by the History heuristic.

Perhaps more importantly, unlike the History heuristic, which requires moves to be sorted based on the values in its arrays, the History-ADS shares the advantageous quality of the Threat-ADS in that it does not require sorting. One can simply explore moves, if applicable, in the order specified by the ADS, thus allowing it to share the strengths of the Killer Moves heuristic, while simultaneously maintaining information on all those moves that have produced a cut.

Lastly, unlike the Threat-ADS, which was specific to the BRS, the History-ADS works within the context of the two-player Mini-Max algorithm. However, since the BRS views a multi-player game as a two-player game by virtue of it treating the opponents as a single entity, the History-ADS heuristic is also applicable to it, and it thus functions in both two-player and multi-player contexts. We will thus be investigating its performance in both these avenues in this work.

6   Refinements to the History-ADS Heuristic

As was discussed above, unlike in the case of the Threat-ADS, which, at most, contained only a few elements that represented the number of opponents, the ADS in the present setting could contain hundreds of elements, as many moves could produce a cut over the course of the search. Thus, it is worthwhile to investigate possible refinements or improvements to the History-ADS heuristic that can potentially mitigate this effect. This section describes two of such possible refinements, which are examined in this paper.

6.1   Bounding the Length of the ADS

Retaining all the information pertaining to moves that have produced a cut is logically beneficial. However, it is possible, and in fact very reasonable to hypothesize, that the majority of savings do not come from moves which are near the tail of the list, but rather near the front. Therefore, if we provide a maximum size on the list, and only retain elements in those positions, it may be possible to noticeably curtail the size of the list, providing some guarantees on its memory performance, while maintaining the vast majority of savings provided by the History-ADS. The way in which we will accomplish this is by forgetting any

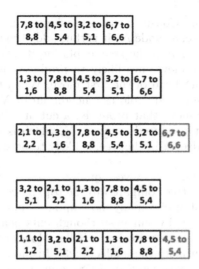

Fig. 5. An example of a history-ADS's list sequence updating over several queries, with a maximum length of 5. The ADS starts with a list of length four, and is queried with the move (1,3) to (1,6), which it moves to the front. It is then queried with (2,1) to (2,2), and as (6,7) to (6,6) is pushed to the sixth position, it is forgotten (highlighted in grey). The process continues as it is queried with (3,2) to (5,1), causing only an internal change, and finally (1,1) to (1,2), pushing (4,5) to (5,4) off the end of the list.

element of the list that falls to position $N + 1$, if the maximum is N. Otherwise, the History-ADS will operate as it was described above. An example of such an ADS updating over several queries, operating with a bounded list, is presented in Fig. 5.

Beyond limiting the memory usage of the History-ADS heuristic, if the developer is attempting to avoid sorting moves by generating them in the order of the ADS, having to traverse a very long list to do this could defeat the purpose of omitting sorting. Thus, demonstrating that the History-ADS can retain the majority of its savings with a smaller list can assist in managing implementation concerns, as well.

6.2   Multi-level ADSs

The History-ADS heuristic as presented earlier maintains a single adaptive list, which is updated whenever a move produces a cut, and is used to order moves when they are encountered elsewhere in the tree. It performs this operation "blindly", without giving consideration to the location in the tree where the move produced a cut, relative to its current location. Thus, if moves are found to produce cuts at the lowest levels of the tree, they will be prioritized at the upper levels of the tree later in the search.

While this may lead to improved savings, as certain moves may be very strong regardless of which level of the tree they occur on, there is a potential weakness in such a blind invocation. Consider the case where a move produces a cut at the highest level of the tree, at node N. It is thus added to the adaptive list, and the search continues deeper into the tree, exploring it in a depth-first manner. Deeper in the tree, many moves are likely to produce cuts, and these will be added to the adaptive list ahead of the first move. When the search returns to the higher levels of the tree, and explores a neighbour of N, these moves will be prioritized first, over the move that produced a cut at its neighbour. However, intuitively the move that produced a cut at N, which is a more similar game state compared to those deeper in the tree, is likely to be stronger at the current node.

Furthermore, by handling all moves equally, as there are many more nodes towards the bottom of the tree compared to the top, moves that are strong near the bottom of the tree will receive many more updates and thus a higher ranking in the adaptive list. This will occur even though cuts near the top of the tree are comparatively more valuable. Both the Killer Moves and History heuristics employ mechanisms to mitigate these effects [23]. Inspired by this, we augment the History-ADS heuristic with multiple ADSs, one for each level, and use them only within the contexts of their sibling nodes.

The use of multi-level ADSs may lead to a reduction in performance, given that learning cannot be applied at different levels of the tree. But given the precedence set by the existing techniques reported in the literature and the potential benefits, we consider it a meaningful avenue of inquiry.

7   Experimental Verification of the Strategy

As the two heuristics share many conceptual commonalities, it is logical to employ a similar set of experiments in analyzing the performance of the History-ADS heuristic that we used in our previous work on the Threat-ADS heuristic. We are interested in learning the improvement gained from using the History-ADS heuristic, when compared to a search that does not employ it. We accomplish this by taking an aggregate of the Node Count (NC) over several turns (where NC is the number of nodes at which computation takes place, omitting those that were pruned), which we then average over fifty trials. We will repeat this experiment with a variety of games, with the Move-to-Front and Transposition update mechanisms, and at varying ply depths, so to provide us with a clear picture of History-ADS, its benefits and drawbacks, and its overall efficacy. While it would seem intuitive to use the runtime as a metric of performance, it has been observed in the literature that CPU time can be a problematic metric for these sorts of experiments, as it is prone to be influenced by the platform used and by the specific implementation [23].

As with our work involving the Threat-ADS heuristic, we will employ the Virus Game, Focus, and Chinese Checkers when considering the multi-player case. However, as we are also considering the two-player case, we require an expanded testing set of games. While the Virus Game, Focus, and Chinese Checkers can all be played with two players, we have elected to also employ some more well-known two-player games, rather than using the same games in their two-player configurations. This is done so as to provide a wider testing base for the History-ADS. The new two-player games that we will employ are Othello, and the very well-known Checkers, or Draughts. The game models are briefly described in the next section.

Since the History-ADS heuristic may be able to retain its knowledge in subsequent turns, we will allow the game to proceed for several turns from the starting configuration. As the Threat-ADS heuristic does not influence the decisions of the BRS, but only its speed of execution, the end result of the game is not a fundamental concern in our experiments. Thus, we will not run the games to termination after this is done. Specifically, we will run the Virus Game for ten turns, Chinese Checkers for five, and Focus for three, which is consistent with our previously presented work. For Othello and Checkers, we allow the game to proceed for five turns in both cases.

As we did for the Threat-ADS heuristic in [15], rather than simply examine the History-ADS heuristic's performance near the start of the game, we also examine its performance in intermediate board states. Compared to the initial configuration, intermediate board states represent a more challenging problem, for a number of reasons. These include a greater degree in the variability of intermediate board positions compared to those close to the start of the game, and the lack of "opening book" knowledge, if applicable, allowing intelligent play to more easily be achieved [8].

During turns within which measurement is taking place, other than the perspective player, all opponents made random moves, to cut down on experiment runtime, as we are interested in tree pruning rather than the final state of the game. However, this is clearly not a valid way to generate intermediate starting board positions, as these would be very unrealistic if only a single player was acting rationally. Thus, when considering the intermediate case, we progress the game initially by having each player use a simple 2-ply alpha-beta search or BRS for a set number of turns, after which we switch to the experimental configuration. The number of turns we advanced into the game in this way was fifteen for the Virus Game, ten for Chinese Checkers and Othello, five for Checkers, and, given its short duration, three for Focus.

In order to determine the statistical significance and impact of the History-ADS heuristic's benefits, we employ the non-parametric Mann-Whitney test to determine the statistical significance, as we do not assume that results follow a normal distribution. We also include the Effect Size measure, to illustrate the size of the effect and as a control against the possibility of over-sampling. In general, an Effect Size of 0.2 is small, 0.5 is medium, and 0.8 is large, with anything substantially larger than that representing an enormous, obvious impact [2].

We first present our results for two-player games, using the standard alpha-beta search technique enhanced with the History-ADS heuristic. We then present those results for the multi-player case, where the History-ADS heuristic is employed to augment the BRS. We then present our results for the refinement proposed in the previous section, that of providing a bound on the size of the adaptive list. Finally, we present our results for our second proposed refinement, namely that of providing an ADS for each level of the tree.

7.1 Game Models

In this section, we will briefly detail the games employed in our experiments, to give the reader a better understanding of their rules and game flow, and to contrast them against each other.

Checkers: Checkers, also known as draughts, is a very well-known board game, designed to be played by two players on an 8x8 checkerboard. During each player's turn, he may move any of his pieces one square diagonally left or right and forward, towards the opponent's side of the board. If the player's piece is adjacent to an opponent's piece, and the square directly across from it is unoccupied, the player may "jump" the opponent's piece, and capture it. If other subsequent jumps are possible, the player can continue to make them, chaining jumps together. If a piece reaches the opposite side of the board, it is promoted, and is no longer restricted to moving only forward. A player loses when his last piece is captured, or when he can make no legal moves. Under normal rules, the game of Checkers requires jumps, if available, to be taken. While this makes the game strategically interesting, it greatly decreases the variability of the game tree's size, and thus we have relaxed this rule, to generate greater search trees. We refer to our variant as Relaxed Checkers. The starting position for Checkers is shown in Fig. 6.

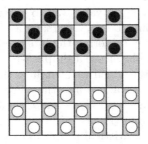

Fig. 6. The starting position for checkers.

Othello: Othello is a two-player board game also played on an 8x8 board, and based on the capture of opponent pieces, although the mechanisms by which capturing takes place are quite different. Initially, each player has two pieces on the board, arranged as in Fig. 7, During his turn, a player may place an additional piece on the board, in a position that "flanks" one or more opponent pieces in a line between the placed piece and another that the player controls. If the player cannot do this, his turn is passed. This captures the enclosed pieces, and they are flipped, or replaced by pieces of the color of the capturing player.

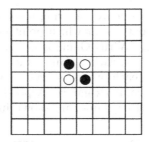

Fig. 7. The starting position for othello.

Play continues until neither player can make a valid move, at which point the player who controls the most pieces is declared the winner.

Focus: Focus is a board game based on piece capturing, designed to be played by two to four players, on an 8×8 board, with the three squares in each corner omitted. The game was originally developed by Sackson in 1969 and released since under many different names [21]. Unlike most other games of its type, Focus allows pieces to be "stacked" on top of each other. The player whose piece is on top of the stack is said to control it, and during his turn, may move one stack he controls, vertically or horizontally, by a number of squares equal to the height of the stack. When a stack is placed on top of another one, they are merged, and all pieces more than five from the top of the stack are removed from the board. If a player captures his own piece, he may place it back on the board in any position, rather than moving a stack. Starting positions for Focus are pre-determined to insure a fair board state. The starting positions for Focus are shown in Fig. 8.

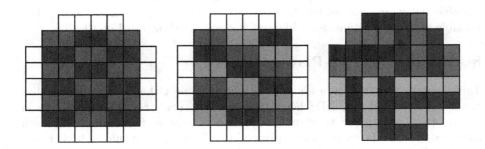

Fig. 8. The two, three, and four player starting positions for Focus.

Virus Game: The Virus Game is a multi-player board game of our own creation, modeled after similar experimental games from previous works, based on a biological metaphor [18]. The Virus Game is, in essence, a "territory control" game where players vie for control of squares on a game board of configurable size (in this work, we use a 5x5 board). During his turn, a player may "infect" a square adjacent to one he controls, at which point he claims that square, and each square adjacent to it. The Virus Game is designed primarily as a highly-configurable testing environment, rather than a tactically interesting game, as it is easy for players to cancel each others' moves; however, it shares many elements in common with more complex games. A possible starting position, and an intermediate state of the Virus Game, are shown in Fig. 9.

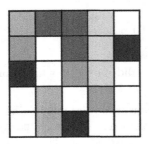 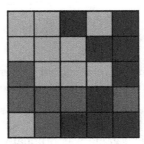

Fig. 9. The virus game at its initial state, and ten turns into the game. Observe that two players have been eliminated, and the pieces are more closely grouped together.

Chinese Checkers: Chinese Checkers is a well-known multi-player board game, played by between two and six players, omitting five players, as it would give one an unfair advantage. The game is played on a star-shaped board, and the objective is to move all of one's pieces to the opposite corner from one's starting position. On his turn, a player may move one of his pieces to one of the adjacent six positions, or "jump" an adjacent piece, which could be either his or his opponent's. As in Checkers, jumps may be chained together as many times as possible, allowing substantial distances to be covered in a single move. The possible starting positions for Chinese Checkers are shown in Fig. 10.

8　Results for Two-Player Games

Table 1 presents our results for the two-player game Othello. We observe that in all cases, the History-ADS heuristic produced very strong improvements in terms of NC, compared to standard alpha-beta search. Furthermore, in each case, the Move-to-Front rule outperformed the Transposition rule. A higher proportion of savings generally correlates with a larger game tree, both in terms of a greater ply depth, and considering the more expansive intermediate case. Our best performance was in the 8-ply intermediate case, with savings of 47 %. We observed an Effect Size ranging between 0.5 and 0.75, indicating a moderate to large effect [2].

Table 2 presents our results for Relaxed Checkers. We notice a very similar trend, compared to Othello, with the History-ADS heuristic generating substantial improvements to pruning in all cases, and generally doing better the larger

Fig. 10. The two, three, four, and six player starting positions for Chinese checkers.

Table 1. Results of applying the History-ADS heuristic for othello in various configurations.

Ply depth	Midgame	Update mechanism	Avg. node count	Std. dev	P-value	Effect size
4	No	None	669	205	-	-
4	No	Move-to-Front	523	162	2.2×10^{-4}	0.71
4	No	Transposition	572	225	0.016	0.47
6	No	None	5061	2385	-	-
6	No	Move-to-Front	3827	1692	6.0×10^{-3}	0.51
6	No	Transposition	4057	2096	0.015	0.42
8	No	None	38,800	20,300	-	-
8	No	Move-to-Front	26,800	12,000	1.2×10^{-3}	0.59
8	No	Transposition	29,700	12,300	0.014	0.45
4	Yes	None	2199	745	-	-
4	Yes	Move-to-Front	1699	633	2.2×10^{-3}	0.67
4	Yes	Transposition	1761	597	0.01	0.59
6	Yes	None	20,100	9899	-	-
6	Yes	Move-to-Front	14,500	6303	6.0×10^{-3}	0.57
6	Yes	Transposition	15,200	6751	0.015	0.49
8	Yes	None	182,000	114,000	-	-
8	Yes	Move-to-Front	95,600	50,200	$<1.0 \times 10^{-5}$	0.76
8	Yes	Transposition	113,000	60,900	1.3×10^{-4}	0.60

the search space is, with Move-to-Front always outperforming Transposition. The best performance was again observed in the 8-ply intermediate case, with a 63 % reduction in tree size, well over half the tree. In this case, the Effect Size ranged between 0.5 and, in cases with less variance, reached levels well over 2 or even 3, suggesting an extreme effect.

Table 2. Results of applying the History-ADS heuristic to Relaxed Checkers in various configurations.

Ply depth	Midgame	Update mechanism	Avg. node count	Std. dev	P-value	Effect size
4	No	None	5930	864	-	-
4	No	Move-to-Front	4712	461	$<1.0 \times 10^{-5}$	1.41
4	No	Transposition	5148	675	$<1.0 \times 10^{-5}$	0.90
6	No	None	78,600	10,600	-	-
6	No	Move-to-Front	40,800	5619	$<1.0 \times 10^{-5}$	3.58
6	No	Transposition	48,600	6553	$<1.0 \times 10^{-5}$	2.84
8	No	None	910,000	172,000	-	-
8	No	Move-to-Front	362,000	55,900	$<1.0 \times 10^{-5}$	3.18
8	No	Transposition	435,000	68,400	$<1.0 \times 10^{-5}$	2.76
4	Yes	None	5447	1859	-	-
4	Yes	Move-to-Front	3772	1257	$<1.0 \times 10^{-5}$	0.90
4	Yes	Transposition	4497	1474	4.5×10^{-3}	0.51
6	Yes	None	64,000	25,700	-	-
6	Yes	Move-to-Front	36,100	12,700	$<1.0 \times 10^{-5}$	1.08
6	Yes	Transposition	43,600	16,600	$<1.0 \times 10^{-5}$	0.79
8	Yes	None	859,000	408,000	-	-
8	Yes	Move-to-Front	317,000	135,000	$<1.0 \times 10^{-5}$	1.33
8	Yes	Transposition	422,000	161,000	$<1.0 \times 10^{-5}$	1.07

Results for our final two-player game, Focus, are shown in Table 3. Yet again, the History-ADS heuristic always produced substantial gains in terms of tree pruning, with larger savings, the Move-to-Front rule always outperformed the Transposition rule, and it did best in larger trees. Our best performance was in the midgame case, with a 77 % reduction in tree size. The Effect Size was over 2 in the more variable intermediate board case, and exceeded 10 in case of an initial board position, again indicating an extreme effect.

9   Results for Multi-player Games

Our results for the multi-player Virus Game are presented in Table 4. We observe very similar behaviour, in comparison to the two-player games. Again, the Move-to-Front rule always outperforms the Transposition rule, and the History-ADS produces substantial gains in all cases, tending towards greater savings in larger trees. The best result was a 55 % reduction in NC, in the 6-ply initial board position case.

Table 3. Results of applying the History-ADS heuristic to two-player Focus in initial and midgame states.

Ply depth	Midgame	Update mechanism	Avg. node count	Std. dev	P-value	Effect size
4	No	None	5,250,000	381,000	-	-
4	No	Move-to-Front	1,290,000	88,000	$<1.0 \times 10^{-5}$	10.39
4	No	Transposition	1,800,000	158,000	$<1.0 \times 10^{-5}$	9.07
4	Yes	None	10,600,000	3,460,000	-	-
4	Yes	Move-to-Front	2,420,000	637,000	$<1.0 \times 10^{-5}$	2.37
4	Yes	Transposition	2,910,000	760,000	$<1.0 \times 10^{-5}$	2.22

Table 4. Results of applying the History-ADS heuristic for the Virus Game in various configurations.

Ply depth	Midgame	Update mechanism	Avg. node count	Std. dev	P-value	Effect size
4	No	None	254,000	28,600	-	-
4	No	Move-to-Front	157,000	17,900	$<1.0 \times 10^{-5}$	3.37
4	No	Transposition	165,000	22,800	$<1.0 \times 10^{-5}$	3.11
6	No	None	10,500,000	1,260,000	-	-
6	No	Move-to-Front	4,690,000	1,010,000	$<1.0 \times 10^{-5}$	4.57
6	No	Transposition	4,850,000	739,000	$<1.0 \times 10^{-5}$	4.45
4	Yes	None	309,000	40,700	-	-
4	Yes	Move-to-Front	188,000	17,800	$<1.0 \times 10^{-5}$	2.97
4	Yes	Transposition	199,000	20,700	$<1.0 \times 10^{-5}$	2.69
6	Yes	None	12,800,000	1,950,000	-	-
6	Yes	Move-to-Front	5,940,000	832,000	$<1.0 \times 10^{-5}$	3.51
6	Yes	Transposition	6,060,000	974,000	$<1.0 \times 10^{-5}$	3.45

Table 5 holds our results for the multi-player variant of Focus. The trends observed are almost identical to the two-player case, and match expectations from patterns recognized there, with a slightly higher maximum of a 78 % reduction in tree size in the intermediate case, given that tree sizes are larger in midgame searches.

Table 5. Results of applying the History-ADS heuristic for the Virus Game in various configurations.

Ply depth	Midgame	Update mechanism	Avg. node count	Std. dev	P-value	Effect size
4	No	None	6,970,000	981,000	-	-
4	No	Move-to-Front	2,180,000	184,000	$<1.0 \times 10^{-5}$	4.88
4	No	Transposition	2,740,000	271,000	$<1.0 \times 10^{-5}$	4.32
4	Yes	None	14,200,000	8,400,000	-	-
4	Yes	Move-to-Front	3,240,000	1,730,000	$<1.0 \times 10^{-5}$	1.30
4	Yes	Transposition	3,570,000	1,860,000	$<1.0 \times 10^{-5}$	1.26

Lastly, Table 6 presents our results for Chinese Checkers, our final multi-player game. Chinese Checkers deviated slightly from established patterns, as performance was very uniform, although the History-ADS heuristic produced large savings in every case, and Move-to-Front continues to outperform Transposition, if slightly in some situations. Our best results were observed in the four player, initial board position case, with a 65 % reduction in tree size, which is, in fact, the smallest search.

Table 6. Results of applying the History-ADS heuristic Chinese Checkers in various configurations.

Ply (Players)	Midgame	Update mechanism	Avg. node count	Std. dev	P-value	Effect size
4 (4-play)	No	None	1,380,000	417,000	-	-
4 (4-play)	No	Move-to-Front	486,000	135,000	$<1.0 \times 10^{-5}$	2.14
4 (4-play)	No	Transposition	505,000	131,000	$<1.0 \times 10^{-5}$	2.09
4 (6-play)	No	None	3,370,000	1,100,000	-	-
4 (6-play)	No	Move-to-Front	1,250,000	316,000	$<1.0 \times 10^{-5}$	1.92
4 (6-play)	No	Transposition	1,320,000	338,000	$<1.0 \times 10^{-5}$	1.87
4 (4-play)	Yes	None	3,340,000	933,000	-	-
4 (4-play)	Yes	Move-to-Front	1,310,000	365,000	$<1.0 \times 10^{-5}$	2.17
4 (4-play)	Yes	Transposition	1,360,000	314,000	$< 1.0 \times 10^{-5}$	2.12
4 (6-play)	Yes	None	8,260,000	2,640,000	-	-
4 (6-play)	Yes	Move-to-Front	3,400,000	899,000	$<1.0 \times 10^{-5}$	1.84
4 (6-play)	Yes	Transposition	3,650,000	920,000	$<1.0 \times 10^{-5}$	1.74

The next section presents our results for the possible refinements to the History-ADS heuristic, described in Sect. 6. Given that, in the case of the History-ADS heuristic, the Move-to-Front rule outperformed the Transposition rule in every single case, sometimes by a large margin, we restrict our update mechanism to it going forward, given its demonstrated superiority.

10   Results for Bounded ADSs

Table 7 presents our results for Othello, with a bounded ADS, in the same configurations as earlier. As is to be expected, due to the History-ADS heuristic being restricted from retaining as much information, some decay in performance was observed, however in even the worst case, the majority of savings were maintained even if the size of the ADS was limited by 5. In a very encouraging scenario, the decrease was by only 1 % (from 26 % to 25 %), when the ply depth was 6 and the size of the list was bounded by 20, from the initial board position.

Table 8 presents our results for Checkers. A similar pattern was observed as in the case of Othello, where the smaller the length of the list, the less the improvement gleaned from the History-ADS heuristic, although the vast majority of savings remained. In the very best case, the reduction in savings was only 2 %, when the size of the list was bounded by 20, from the initial board state (regardless of ply depth).

Consider Table 9, which presents our results for Focus with varying limits on the ADS' size. We observe the same pattern as we did in Othello and Checkers, in both the two-player and the multi-player cases, although the difference in

Table 7. Results of applying the History-ADS heuristic to Othello with a varying maximum length on the ADS.

Ply depth	Midgame	Limit	Avg. node count	Std. dev	P-value	Effect size
4	No	No ADS	669	205	-	-
4	No	Unbound	525	175	3.2×10^{-4}	0.70
4	No	20	572	196	5.8×10^{-3}	0.47
4	No	5	596	196	0.041	0.36
6	No	No ADS	5061	2385	-	-
6	No	Unbound	3727	1552	1.7×10^{-3}	0.56
6	No	20	3779	1884	3.3×10^{-3}	0.54
6	No	5	3961	1603	0.013	0.46
8	No	No ADS	38,800	20,300	-	-
8	No	Unbound	23,600	10,800	$<1.0 \times 10^{-5}$	0.75
8	No	20	24,900	11,000	1.1×10^{-5}	0.68
8	No	5	28,600	12,200	5.8×10^{-3}	0.50
4	Yes	No ADS	2199	745	-	-
4	Yes	Unbound	1702	570	2.7×10^{-4}	0.67
4	Yes	20	1787	663	2.6×10^{-3}	0.55
4	Yes	5	1840	576	0.010	0.48
6	Yes	No ADS	20,100	9899	-	-
6	Yes	Unbound	13,300	6916	7.0×10^{-5}	0.69
6	Yes	20	13,900	6846	6.9×10^{-4}	0.63
6	Yes	5	14,800	7916	1.7×10^{-3}	0.54
8	Yes	No ADS	182,000	114,000	-	-
8	Yes	Unbound	92,900	49,700	$<1.0 \times 10^{-5}$	0.79
8	Yes	20	109,000	63,800	7.0×10^{-5}	0.65
8	Yes	5	116,000	59,800	3.1×10^{-5}	0.58

savings between an unbounded list and a list of length 20 or 5 is quite a bit smaller, with even the worst case being 78 % to 75 % in the 4-ply midgame case, which was only a 3 % difference.

Table 10 contains our results for the Virus Game. Our observations were analogous to those for the other three cases presented so far. While the History-ADS heuristic produced noticeable gains in all cases, they were lessened by a limit being placed on the maximum length of the ADS. In the very worst case, the change observed was a reduction in savings from 56 % to 51 % in the 6-ply case, with measurements taken from the initial board state, a very small change of only 5 %.

Finally, Table 11 presents our results for Chinese Checkers. The patterns we observed were similar to those for the other four games, however in the case of four-player Chinese Checkers, from the initial board position, a maximum list size of 5 outperformed a list size of 20. As before, for this game, variability was

Table 8. Results of applying the History-ADS heuristic to Relaxed Checkers with a varying maximum length on the ADS.

Ply depth	Midgame	Limit	Avg. node count	Std. dev	P-value	Effect size
4	No	No ADS	5930	864	-	-
4	No	Unbound	4638	493	$<1.0 \times 10^{-5}$	1.49
4	No	20	4755	406	$<1.0 \times 10^{-5}$	1.36
4	No	5	4776	516	$<1.0 \times 10^{-5}$	1.33
6	No	No ADS	78,600	10,600	-	-
6	No	Unbound	41,000	5588	$<1.0 \times 10^{-5}$	3.55
6	No	20	42,700	5292	$<1.0 \times 10^{-5}$	3.39
6	No	5	44,700	5545	$<1.0 \times 10^{-5}$	3.20
8	No	No ADS	910,000	172,000	-	-
8	No	Unbound	358,000	53,200	$<1.0 \times 10^{-5}$	3.20
8	No	20	375,000	55,700	$<1.0 \times 10^{-5}$	3.11
8	No	5	398,000	78,300	$<1.0 \times 10^{-5}$	2.97
4	Yes	No ADS	5447	1859	-	-
4	Yes	Unbound	3681	1138	$<1.0 \times 10^{-5}$	0.94
4	Yes	20	4130	1304	1.9×10^{-4}	0.70
4	Yes	5	4217	1363	4.3×10^{-4}	0.66
6	Yes	No ADS	64,000	25,700	-	-
6	Yes	Unbound	34,400	12,400	$<1.0 \times 10^{-5}$	1.15
6	Yes	20	36,700	14,200	$<1.0 \times 10^{-5}$	1.06
6	Yes	5	39,500	16,800	$<1.0 \times 10^{-5}$	0.96
8	Yes	No ADS	859,000	408,000	-	-
8	Yes	Unbound	293,000	100,000	$<1.0 \times 10^{-5}$	1.39
8	Yes	20	333,000	133,000	$<1.0 \times 10^{-5}$	1.29
8	Yes	5	397,000	193,000	$<1.0 \times 10^{-5}$	1.13

quite small, such as 62 % to 61 % in the 4-ply, six player case, with a limit on the list of size of 20. Considering even the worst situation, the largest change observed was 66 % to 62 %, in the 4-ply, four player case, with measurements taken from the initial board position.

11   Results for Multi-level ADSs

Table 12 presents our results for multi-level ADSs, in the domain of Othello. We observed that, similar to the limit on the ADS, the use of a multi-level ADS reduces performance by a consistent but small amount. Limiting the multi-level ADS's size further reduces the performance, in the majority of cases. In the best case, savings were reduced from 39 % to 36 %, when a multi-level ADS with no size limitation was employed, in the 8-ply case, from the initial board position.

Our results for Relaxed Checkers are shown in Table 13. The patterns observed were similar to those for Othello, with the use of multi-level ADSs

Table 9. Results of applying the History-ADS heuristic to Focus with a varying maximum length on the ADS.

Ply depth	Midgame	Limit	Avg. node count	Std. dev	P-value	Effect size
4	No	No ADS	5,250,000	381,000	-	-
4	No	Unbound	1,260,000	90,900	$<1.0 \times 10^{-5}$	10.46
4	No	20	1,270,000	96,300	$<1.0 \times 10^{-5}$	10.43
4	No	5	1,330,000	108,000	$<1.0 \times 10^{-5}$	10.28
4 (Multi)	No	No ADS	6,970,000	981,000	-	-
4 (Multi)	No	Unbound	2,150,000	165,000	$<1.0 \times 10^{-5}$	4.92
4 (Multi)	No	20	2,210,000	165,000	$<1.0 \times 10^{-5}$	4.86
4 (Multi)	No	5	2,230,000	154,000	$<1.0 \times 10^{-5}$	4.79
4	Yes	No ADS	10,600,000	3,460,000	-	-
4	Yes	Unbound	2,390,000	631,000	$<1.0 \times 10^{-5}$	2.37
4	Yes	20	2,520,000	710,000	$<1.0 \times 10^{-5}$	2.34
4	Yes	5	2,630,000	847,000	$<1.0 \times 10^{-5}$	2.30
4 (Multi)	Yes	No ADS	14,200,000	8,400,000	-	-
4 (Multi)	Yes	Unbound	3,120,000	1,700,000	$<1.0 \times 10^{-5}$	1.31
4 (Multi)	Yes	20	3,310,000	1,700,000	$<1.0 \times 10^{-5}$	1.30
4 (Multi)	Yes	5	3,370,000	1,410,000	$<1.0 \times 10^{-5}$	1.29

Table 10. Results of applying the History-ADS heuristic to the Virus Game with a varying maximum length on the ADS.

Ply depth	Midgame	Limit	Avg. node count	Std. dev	P-value	Effect size
4	No	No ADS	254,000	28,600	-	-
4	No	Unbound	154,000	20,800	$<1.0 \times 10^{-5}$	3.47
4	No	20	158,000	19,000	$<1.0 \times 10^{-5}$	3.36
4	No	5	163,000	19,000	$<1.0 \times 10^{-5}$	3.20
6	No	No ADS	10,500,000	1,260,000	-	-
6	No	Unbound	4,650,000	767,000	$<1.0 \times 10^{-5}$	4.60
6	No	20	4,830,000	640,000	$<1.0 \times 10^{-5}$	4.46
6	No	5	5,110,000	658,000	$<1.0 \times 10^{-5}$	4.24
4	Yes	No ADS	308,000	40,700	-	-
4	Yes	Unbound	187,000	23,100	$<1.0 \times 10^{-5}$	3.00
4	Yes	20	187,000	22,200	$<1.0 \times 10^{-5}$	3.00
4	Yes	5	194,000	21,600	$<1.0 \times 10^{-5}$	2.81
6	Yes	No ADS	12,800,000	1,950,000	-	-
6	Yes	Unbound	5,870,000	863,000	$<1.0 \times 10^{-5}$	3.55
6	Yes	20	5,910,000	821,000	$<1.0 \times 10^{-5}$	3.53
6	Yes	5	6,390,000	767,000	$<1.0 \times 10^{-5}$	3.29

Table 11. Results of applying the History-ADS heuristic to Chinese checkers with a varying maximum length on the ADS.

Ply depth	Midgame	Limit	Avg. node count	Std. dev	P-value	Effect size
4 (4-play)	No	No ADS	1,380,000	417,000	-	-
4 (4-play)	No	Unbound	476,000	125,000	$<1.0 \times 10^{-5}$	2.16
4 (4-play)	No	20	531,000	125,000	$<1.0 \times 10^{-5}$	2.03
4 (4-play)	No	5	525,000	157,000	$<1.0 \times 10^{-5}$	2.04
4 (6-play)	No	No ADS	3,370,000	1,100,000	-	-
4 (6-play)	No	Unbound	1,280,000	368,000	$<1.0 \times 10^{-5}$	1.90
4 (6-play)	No	20	1,330,000	332,000	$<1.0 \times 10^{-5}$	1.85
4 (6-play)	No	5	1,360,000	297,000	$<1.0 \times 10^{-5}$	1.83
4 (4-play)	Yes	No ADS	3,340,000	933,000	-	-
4 (4-play)	Yes	Unbound	1,240,000	335,000	$<1.0 \times 10^{-5}$	2.25
4 (4-play)	Yes	20	1,330,000	330,000	$<1.0 \times 10^{-5}$	2.16
4 (4-play)	Yes	5	1,370,000	455,000	$<1.0 \times 10^{-5}$	2.11
4 (6-play)	Yes	No ADS	8,260,000	1,950,000	-	-
4 (6-play)	Yes	Unbound	3,200,000	863,000	$<1.0 \times 10^{-5}$	1.92
4 (6-play)	Yes	20	3,290,000	821,000	$<1.0 \times 10^{-5}$	1.88
4 (6-play)	Yes	5	3,340,000	767,000	$<1.0 \times 10^{-5}$	1.86

Table 12. Results of applying the History-ADS heuristic to Othello with multi-level ADSs.

Ply depth	Midgame	Limit	Avg. node count	Std. dev	P-value	Effect size
6	No	No ADS	5061	2385	-	-
6	No	No	3727	1552	1.7×10^{-3}	0.56
6	No	Yes	4110	1828	0.023	0.40
6	No	Yes (5-limit)	4305	1843	0.089	0.37
8	No	No ADS	38,800	20,300	-	-
8	No	No	23,600	10,800	$<1.0 \times 10^{-5}$	0.75
8	No	Yes	24,900	11,900	1.0×10^{-4}	0.68
8	No	Yes (5-limit)	25,600	13,100	2.9×10^{-4}	0.65
6	Yes	No ADS	20,100	9899	-	-
6	Yes	No	13,300	6916	7.0×10^{-5}	0.69
6	Yes	Yes	14,700	7279	1.8×10^{-3}	0.55
6	Yes	Yes (5-limit)	16,000	7283	0.017	0.42
8	Yes	No ADS	182,000	114,000	-	-
8	Yes	No	92,900	49,700	$<1.0 \times 10^{-5}$	0.79
8	Yes	Yes	110,000	56,500	1.0×10^{-4}	0.64
8	Yes	Yes (5-limit)	105,000	50,600	$<1.0 \times 10^{-5}$	0.68

Table 13. Results of applying the History-ADS heuristic to Relaxed Checkers with multi-level ADSs.

Ply depth	Midgame	Limit	Avg. node count	Std. dev	P-value	Effect size
6	No	No ADS	78,600	10,600	-	-
6	No	No	41,000	5588	$<1.0 \times 10^{-5}$	3.55
6	No	Yes	45,300	5977	$<1.0 \times 10^{-5}$	3.14
6	No	Yes (5-limit)	46,900	6727	$<1.0 \times 10^{-5}$	3.00
8	No	No ADS	910,000	172,000	-	-
8	No	No	358,000	53,200	$<1.0 \times 10^{-5}$	3.20
8	No	Yes	394,000	53,500	$<1.0 \times 10^{-5}$	2.99
8	No	Yes (5-limit)	416,000	68,300	$<1.0 \times 10^{-5}$	2.86
6	Yes	No ADS	64,000	25,700	-	-
6	Yes	No	34,400	12,400	$<1.0 \times 10^{-5}$	1.15
6	Yes	Yes	38,100	15,400	$<1.0 \times 10^{-5}$	1.06
6	Yes	Yes (5-limit)	37,800	13,000	$<1.0 \times 10^{-5}$	1.02
8	Yes	No ADS	859,000	408,000	-	-
8	Yes	No	293,000	100,000	$<1.0 \times 10^{-5}$	1.39
8	Yes	Yes	350,000	163,000	$<1.0 \times 10^{-5}$	1.25
8	Yes	Yes (5-limit)	404,000	150,000	$<1.0 \times 10^{-5}$	1.11

performing slightly worse than the original version. The smallest change was from 46 % to 40 %, in the 6-ply case, where measurements were taken from a midgame board position.

Lastly, we present our results for the Virus Game with multi-level ADSs in Table 14. While we observed, as before, that the performance worsened when the multi-level approach was employed, the difference was quite a bit smaller in the context of the Virus Game. The loss in savings was very slight in the best case, from 56 % to 55 %, from the initial board position.

Table 14. Results of applying the History-ADS heuristic to Focus with multi-level ADSs.

Ply depth	Midgame	Limit	Avg. node count	Std. dev	P-value	Effect size
6	No	No ADS	10,500,000	1,260,000	-	-
6	No	No	4,650,000	767,000	$<1.0 \times 10^{-5}$	4.60
6	No	Yes	4,740,000	598,000	$<1.0 \times 10^{-5}$	4.53
6	No	Yes (5-limit)	4,800,000	817,000	$<1.0 \times 10^{-5}$	4.49
6	Yes	No ADS	12,800,000	1,950,000	-	-
6	Yes	No	5,870,000	863,000	$<1.0 \times 10^{-5}$	3.55
6	Yes	Yes	5,990,000	653,000	$<1.0 \times 10^{-5}$	3.49
6	Yes	Yes (5-limit)	6,160,000	616,000	$<1.0 \times 10^{-5}$	3.40

12   Discussion

The results presented in the previous sections strongly reinforce the hypothesis that an ADS managing move history, employed by the History-ADS heuristic, can achieve improvements in tree pruning through better move ordering, in both two-player and multi-player games. This confirms that such an ADS does, indeed, correctly prioritize the most effective moves, based on their previous performance elsewhere in the tree, as initially hypothesized.

Our results confirm that the History-ADS heuristic is able to achieve a statistically significant reduction in NC in three two-player games, Othello, Relaxed Checkers, and the two-player variant of Focus, as well as three multi-player games, the Virus Game, Chinese Checkers, and the multi-player variant of Focus. Compared to our previous work on the Threat-ADS heuristic, where the degree of reduction ranged between 5 % and 20 %, and which centered around 10 %, we observe a much more drastic improvement in pruning when the History-ADS is employed, to a value as high as 78 %. The History-ADS heuristic displayed quite variable performance, with higher savings generally correlated with the size of the tree. In the case of ply depth, this is intuitively appealing, as cuts made earlier higher in the tree can lead to large numbers of moves being pruned, and if branching factor is higher, as in Focus, many moves are unlikely to be particularly strong, and so correct move ordering could, very reasonably, have a high impact on performance.

Our second observation is that in all cases, the Move-to-Front rule outperformed the Transposition rule. This is a reasonable outcome, as unlike with the Threat-ADS, the adaptive list may contain dozens to hundreds of elements, and it would take quite a bit of time for the Transposition rule to migrate a particularly strong move to the head of the list. As opposed to this, the Move-to-Front would migrate the move to the front quickly, and would likely keep it there. This phenomenon also confirms that the order of elements within the list matters to the move ordering. In other words, merely maintaining an unsorted collection of moves that have produced a cut will not perform as well as employing an adaptive list, further supporting the use of the History-ADS heuristic.

The exception to the above trend occurs in the case of Chinese Checkers. While Chinese Checkers, with its large branching factor, usually sees a larger reduction in tree size compared to Othello or Relaxed Checkers, savings for all the cases within the context of Chinese Checkers are roughly equivalent. Furthermore, despite having the smallest overall tree size, we observe that the best performance occurs for the four player case, with measurements being taken from the initial board position. What this suggests is that moves that produce a cut in Chinese Checkers may not have a very strong natural ranking between them, and so attempting to rank them in the ADS does not help as much as it does for the other games. This would also explain why the Move-to-Front rule did not outperform the Transposition rule to the same extent as in the other cases.

We found that when limiting the maximum size of the ADS, while there was some reduction in performance, as was expected, the loss was very slight in most cases. This confirms our hypothesis that elements near the head of the list tend to remain there, and provide the majority of the move ordering benefits, with diminishing returns as the list gets longer. The fact that the majority of savings are still maintained in all cases even when the list is limited to only contain five elements, successfully addresses one of the concerns we had with of the History-ADS heuristic discussed earlier, namely that the length of the adaptive list can potentially be quite large.

The multi-level ADS approach was found to do worse than the single ADS approach, suggesting that savings that may be gained for prioritizing moves that produced a cut on the current level of the tree, if they exist, are offset by the inability to apply what the algorithm has learned to other levels of the tree. We can thus conclude that the absolutist approach of having a separate ADS at each level of the tree is likely not the optimal way to address concerns of overvaluing moves that are strong near the bottom of the tree. However, the multi-level ADS approach may not be completely useless. If the History-ADS heuristic is employed alongside other, perhaps domain-specific move ordering heuristics, then prioritizing only the best moves at each level may be as effective. This approach begs for more investigation.

Despite the presence of multiple ADSs, limiting the size of the list to 5 reduced improvements even more. Observing the internal functioning of the search, the reason for this appears to be that the limit is only a factor at the lowest levels of the tree, where many more moves produce cuts and the limit impacts the ADS heavily. As opposed to this, levels closer to the root do not require as much space. This is especially visible in the case of 8-ply Othello from the midgame case, where a multi-level ADS with a maximum size of 5 did not, in fact, produce a statistically significant improvement in tree pruning, which is the first time that such an event has occurred for the History-ADS heuristic. Overall, however, the multi-level ADS' performance was close to the original version in the vast majority of cases.

13 Conclusions, Contributions and Future Work

In this work, we introduced the concept of ordering moves, in the alpha-beta search algorithm, based on an ADS, which is dynamically reorganized according to its update mechanism that ranks moves based on which moves have performed well earlier in the search. We have named this technique the History-ADS heuristic. Our results demonstrate conclusively that the History-ADS heuristic is able to obtain a substantial reduction in tree size, in a range of two-player and multi-player games. Specifically, its efficacy has been proven in Othello, Checkers, Focus, Chinese Checkers, and the Virus Game of our own invention, which represents a wide variety of games with different rules and strategies. The results strongly support the hypothesis that the History-ADS heuristic can perform well in the context of two-player and multi-player games.

Despite the power of the History-ADS heuristic, we observed that it retains a particularly large list of moves. We conjecture that the majority of those moves, especially near the tail of the list, may be pruned to save space and assist in implementation, without much loss of performance. Our results strongly support this hypothesis, and we found that in many games, even in the worst case, only a small proportion of savings was lost when restricting the formerly unlimited list to a size of only five, thus overcoming one of the History-ADS heuristic's main drawbacks.

While the use of an ADS at each level of the tree performed similarly to the single ADS, it was outperformed by the single ADS in all cases. However, given the strong basis in the literature of techniques that consider moves based on the level of the tree where they are found, such as with the Killer Moves and History heuristics, we believe that it is worthwhile to investigate this area further. In the hope of striking a more reasonable balance between the two extremes, work is currently ongoing on methods to prioritize learning at the level of the tree where it was acquired, while not completely excluding information obtained elsewhere.

The History-ADS heuristic serves as a natural expansion upon our previous work, introducing the Threat-ADS heuristic, and demonstrates that the use of ADSs in the context of game playing has applicability outside of the window of ranking opponents based on their threats. This work provides an even stronger basis for further examination of the applicability of ADS-based techniques to game playing, and we hope it will inspire others to examine these possibilities in the future.

References

1. Albers, S., Westbrook, J.: Self-organizing data structures. In: Fiat, A. (ed.) Online Algorithms 1996. LNCS, vol. 1442, pp. 13–51. Springer, Heidelberg (1998)
2. Coe, R.: It's the effect size, stupid: what effect size is and why it is important. In: Annual Conference of the British Educational Research Association, University of Exeter, Exeter, Devon (2002)
3. Corman, T.H., Leiserson, C.E., Rivest, R.L., Stein, C.: Introduction to Algorithms, 3rd edn, pp. 302–320. MIT Press, Upper Saddle River (2009)
4. Estivill-Castro, V.: Move-to-end is best for double-linked lists. In: Proceedings of the Fourth International Conference on Computing and Information, pp. 84–87 (1992)
5. Gonnet, G.H., Munro, J.I., Suwanda, H.: Towards self-organizing linear search. In: Proceedings of the Annual Symposium on Foundations of Computer Science (FOCS 1979), pp. 169–171 (1979)
6. Hester, J.H., Hirschberg, D.S.: Self-organizing linear search. ACM Comput. Surv. **17**, 285–311 (1985)
7. Knuth, D.E., Moore, R.W.: An analysis of alpha-beta pruning. Artifi. Intell. **6**, 293–326 (1975)
8. Levene, M., Bar-Ilan, J.: Comparing typical opening move choices made by humans and chess engines. Computing Research Repository (2006)
9. Luckhardt, C., Irani, K.: An algorithmic solution of n-person games. In: Proceedings of the AAAI 1986, pp. 158–162 (1986)
10. Papadoupoulus, A.: Exploring optimization strategies in board game abalone for alpha-beta search. In: Proceedings of the 2012 IEEE Conference on Computational Intelligence and Games (CIG 2012), pp. 63–70 (2012)

11. Pettie, S.: Splay trees, davenport-schinzel sequences, and the deque conjecture. In: Proceedings of the Nineteenth Annual ACM-SIAM Symposium on Discrete Algorithms (2008)
12. Polk, S., Oommen, B.J.: On applying adaptive data structures to multi-player game playing. In: Bramer, M., Petridis, M. (eds.) Research and Development in Intelligent Systems XXX, pp. 125–138. Springer, Heidelberg (2013)
13. Polk, S., Oommen, B.J.: On enhancing recent multi-player game playing strategies using a spectrum of adaptive data structures. In: Proceedings of the 2013 Conference on Technologies and Applications of Artificial Intelligence (TAAI 2013) (2013)
14. Polk, S., Oommen, B.J.: Enhancing history-based move ordering in game playing using adaptive data structures. In: Núñez, M., Nguyen, N.T., Camacho, D., Trawiński, B. (eds.) ICCCI 2015. LNCS, vol. 9329, pp. 225–235. Springer, Heidelberg (2015). doi:10.1007/978-3-319-24069-5_21
15. Polk, S., Oommen, B.J.: Novel AI strategies for multi-player games at intermediate board states. In: Ali, M., Kwon, Y.S., Lee, C.-H., Kim, J., Kim, Y. (eds.) IEA/AIE 2015. LNCS, vol. 9101, pp. 33–42. Springer, Heidelberg (2015)
16. Polk, S., Oommen, B.J.: Space and depth-related enhancements of the history-ADS strategy in game playing. In: Proceedings of the 2015 IEEE Conference on Computational Intelligence and Games (CIG 2015) (2015)
17. Reinefeld, A., Marsland, T.A.: Enhanced iterative-deepening search. IEEE Trans. Pattern Anal. Mach. Intell. **16**, 701–710 (1994)
18. Rendell, P.: A universal turing machine in conway's game of life. In: Proceedings of the International Conference on High Performance Computing and Simulation (HPCS 2011), pp. 764–772 (2011)
19. Rivest, R.L.: On self-organizing sequential search heuristics. In: Proceedings of the IEEE Symposium on Switching and Automata Theory, pp. 63–67 (1974)
20. Russell, S.J., Norvig, P.: Artificial Intelligence: A Modern Approach, 3rd edn. Prentice-Hall Inc., Upper Saddle River (2009)
21. Sacksin, S.: A Gamut of Games. Random House, New York (1969)
22. Schadd, M.P.D., Winands, M.H.M.: Best reply search for multiplayer games. IEEE Trans. Comput. Intell. AI Games **3**, 57–66 (2011)
23. Schaeffer, J.: The history heuristic and alpha-beta search enhancements in practice. IEEE Trans. Pattern Anal. Mach. Intell. **11**, 1203–1212 (1989)
24. Schrder, E.: Move ordering in rebel. Discussion of move ordering techniques used in REBEL, a powerful chess engine (2007)
25. Shannon, C.E.: Programming a computer for playing Chess. Phil. Mag. **41**, 256–275 (1950)
26. Sleator, D.D., Tarjan, R.E.: Amortized efficiency of list update and paging rules. Commun. ACM **28**, 202–208 (1985)
27. Sturtevant, N.R.: A comparison of algorithms for multi-player games. In: Schaeffer, J., Müller, M., Björnsson, Y. (eds.) CG 2002. LNCS, vol. 2883, pp. 108–122. Springer, Heidelberg (2003)
28. Sturtevant, N., Games, M.-P.: Algorithms and Approaches. Ph.D. thesis, University of California (2003)
29. Sturtevant, N., Bowling, M.: Robust game play against unknown opponents. In: Proceedings of the International Joint Conference on Autonomous Agents and Multiagent Systems (AAMAS 2006), pp. 713–719 (2006)
30. Sturtevant, N., Zinkevich, M., Bowling, M.: Prob-Maxn: playing n-player games with opponent models. In: Proceedings of the National Conference on Artificial Intelligence (AAAI 2006), pp. 1057–1063 (2006)

31. Szita, I., Chaslot, G., Spronck, P.: Monte-carlo tree search in settlers of catan. In: van den Herik, H.J., Spronck, P. (eds.) ACG 2009. LNCS, vol. 6048, pp. 21–32. Springer, Heidelberg (2010)

32. Zuckerman, I., Felner, A., Kraus, S.: Mixing search strategies for multi-player games. In Proceedings of the Twenty-first International Joint Conferences on Artificial Intelligence (IJCAI 2009), pp. 646–651 (2009)

Identification of Possible Attack Attempts Against Web Applications Utilizing Collective Assessment of Suspicious Requests

Marek Zachara[⊠]

AGH University of Science and Technology, Kraków, Poland
mzachara@agh.edu.pl

Abstract. The number of web-based activities and websites is growing every day. Unfortunately, so is cyber-crime. Every day, new vulnerabilities are reported and the number of automated attacks is constantly rising. In this article, a new method for detecting such attacks is proposed, whereas cooperating systems analyze incoming requests, identify potential threats and present them to other peers. Each host can then utilize the knowledge and findings of the other peers to identify harmful requests, making the whole system of cooperating servers "remember" and share information about the existing threats, effectively "immunizing" it against them.

The method was tested using data from seven different web servers, consisting of over three million of recorded requests. The paper also includes proposed means for maintaining the confidentiality of the exchanged data and analyzes impact of various parameters, including the number of peers participating in the exchange of data. Samples of identified attacks and most common attack vectors are also presented in the paper.

Keywords: Websites · Applications security · Threat detection · Collective decision

1 Introduction

According to a recent report by Netracft [14], the number of websites around the world is estimated to be almost 850 million, with 150 million of them considered "active". People rely on Internet and the websites in their daily activities, trusting them with their data and their money.

Unfortunately, with more and more data and resources handled by websites, they have become an attractive prey to criminals, both individuals and organized crime. It is very difficult to measure the scale of the cyber-threats, and their impact on the companies and the economy as a whole, since there is no commonly accepted methodology available yet. As an example, McAfee estimates that the cost of the cybercrime reaches 1.5 % of the GDP for the Netherlands and Germany [12]. There is also a more detailed study for the UK [2]. However, such estimations might be imprecise, because they are based on imperfect surveys, which may lead to a high estimation error, as explained in [4,8].

© Springer-Verlag Berlin Heidelberg 2016
N.T. Nguyen and R. Kowalczyk (Eds.): TCCI XXII, LNCS 9655, pp. 45–59, 2016.
DOI: 10.1007/978-3-662-49619-0_3

1.1   Vulnerability of Web Applications

The initial web sites in the 1990 s were meant primarily for publishing and dissemination of information. Virtually all of their resources were meant for public access. The HTTP protocol developed then, and still used today for the transport of the web pages, does not even include any means of tracking or controlling user sessions. Today's web sites are, however, quite different. They often gather and control valuable data - including personal data, passwords or bank accounts.

Symantec claims that while running a thousand of vulnerability scans per day, they found approximately 76 % of the scanned websites to have at least one unpatched vulnerability, with 20 % of the servers having critical vulnerabilities [21]. In another report [22], WhiteHat Security stated that 86 % of the web applications they tested had at least one serious vulnerability, with an average of 56 vulnerabilities per web application.

It can be assumed, that one of the reasons for the low security of web applications is their uniqueness. While the underlying operating systems, web servers, firewalls, and databases are usually well known and tested products, that are subject to continuous scrutiny by thousands of users, a web application is often on its own, with its security depending primarily on the owner and developers' skills and will.

1.2   Malware and Automated Attacks

The massive amount of websites, and their availability over the Internet, led to a rise of automated methods and tools for scanning and possibly breaking into them. Bot-nets and other malware are often targeting websites for known vulnerabilities. For example, Symantec in one of their previous reports stated that in just the single month of May in 2012 the LizaMoon toolkit was responsible for at least a million successful SQL Injections attacks, and that approximately 63 % of websites used to distribute malware were actually legitimate websites compromised by attackers.

Easy access to information and ready-made tools for scanning and exploitation of websites' vulnerabilities resulted in a large number of individuals, collectively known as "script kiddies" attempting random break-in attempts against them, using the same tools downloaded from the Internet.

1.3   The Tools for Battling the Attacks

The reason why firewalls do not protect websites from harmful requests is that their ability is only to filter traffic at the lower layers of the OSI model (usually up to layer 5 for stateful firewalls), while the identification of harmful requests is only possible at layer 7. There are specialized firewalls, known as Web Application Firewalls (WAF) [15,16], but their adoption is limited, primarily because of the time and cost required to configure and maintain them.

There are two broad classes of methods employed for battling attack attempts and identification of the harmful data arriving at the server. The first group consists of various signature-related methods, similar to the popular anti-virus software. Some examples are provided in [7,19]. These methods try to identify known malicious attack patterns on the basis of their knowledge (provided a priori). The primary benefit of such methods is the low number of false-positive alarms. Their primary drawback is the inability to identify new threats and new attack vectors, until they are evaluated by some entity (e.g. a security expert), and introduced into the knowledge database. Within the fast-changing Internet threat environment, they provide very limited protection against attackers, as vulnerabilities are exploited often within hours of their disclosure (citing the Symantec report again [21]: "Within four hours of the Heartbleed vulnerability becoming public in 2014, Symantec saw a surge of attackers stepping up to exploit it").

The second group of methods relies on various types of heuristics to identify potential threats in real time. This approach has been employed for network traffic analysis and intrusion detection [5,17] with SNORT [18] being the well-known open source implementation of such Intrusion Detection Systems (IDS). Although there have been a number of attempts to utilize similar approach in order to secure websites [3,11], these methods usually rely on very simple heuristics and a simple decision tree. There is also a method developed by the author of this article that utilizes weighted graph for modeling and storage of users' typical page-traversing paths that has been presented in [23].

1.4   Rationale for the Collective Assessment

All these methods previously described rely on local evaluation and assessment of the requests, utilizing the knowledge either provided or acquired at a single web server. However, the rise of automated tools and malware led to a situation where the same attack vectors, or even the same requests are used to scan and attack various unrelated websites. Establishing a method of sharing the information between web servers about encountered malicious requests could therefore provide substantial benefits in protecting the websites against these attacks.

2   The Principles of the Method

The attacks on web applications/web servers are usually done either by manipulating parameters sent with a request (*parameter tampering*) or by requesting URLs different to these expected by the application (*forceful browsing*). A good account on specific attacks and their impact can be found in [6]. An example of such attacks are presented in Fig. 1, where the attacker is apparently trying to identify, at various possible locations, a presence of phpMyAdmin - a commonly used web-based database management module.

The web server's log files are usually the easiest way to retrieve information about the requests arriving at the server, allowing for easy parsing and identification of suspicious requests. They have been widely used for this purpose [1,9], and the reference implementation of the proposed method also utilizes the log files as the source of information.

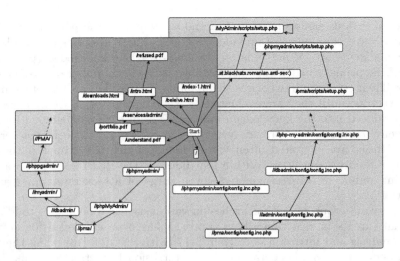

Fig. 1. Sample users' sessions extracted from a web server's log file, with attack attempts marked by red background (Color figure online)

A high-level overview of the proposed method is presented in Fig. 2. As can be seen there, each server maintains its own list of suspicious requests, which is published for other servers to download. On the other hand, the server retrieves similar lists from other peers and keep cached copies of them. Each new request arriving at the server is evaluated against these lists, which constitute the server's knowledge about the currently observed attack patterns. Requests that are considered as potentially harmful are then reported to the administrator.

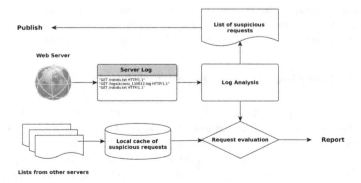

Fig. 2. Overview of the data flow for the collective identification process.

2.1   The Key Concepts of the Proposed Method

This method relies on the fact that most attacks are automated. They are performed either by malware (like LizaMoon mentioned before), or "script kiddies" who download and use ready-made attack tool-kits. These two groups have the ability to perform attacks attempts on a large scale. However, these attacks are usually identical, or very similar to each other. Since they are performed by programs and usually rely on one or just a few attack vectors, they generate similar requests to a large number of web servers.

For clarity reasons, let's reiterate the features of the attacks performed by malware and "script kiddies":

- **Distributed.** The attacks are usually targeted towards a number of web hosts, not at the same time, but within a limited time-frame.
- **Similar.** Malware and "script kiddies" both use single pieces of software with incorporated attack vectors. This software produces identical or very similar requests to various hosts.
- **Unusual.** The requests generated by the automated software are not tailored for the specific host, so they often do not match the host's application.

Fortunately, these features can be utilized to detect and neutralize a major part of the threats coming from these sources. These attacks target a large number of hosts (web applications) in hope of finding a few that will be vulnerable to the specific attack vector. In the process however, they send requests to hosts that are either not vulnerable, or in most cases - do not have the specific module installed. Such requests can easily be identified at these hosts as suspicious and presented to other hosts as an example of an abnormal behavior. Other hosts, which retrieve the information, are then aware that certain requests have been made to a number of other hosts and were considered suspicious by them. This knowledge, in turn, can be utilized to assess and possibly report an incoming requests. Specific methods of such assessment will be discussed later, but it is worth to note that a very similar mode of operation is used by our immune system, where antigens of an infection are presented by body cells to the T-cells, which in turn coordinate the response of the system by passing the 'knowledge' about the intruder to the other elements of the immune system.

2.2   The Exchange of Information

There are various means that can be employed to facilitate the exchange of information between web servers about suspicious attacks. Large scale service providers can opt for a private, encrypted channels of communication, but the general security of the web would benefit from a public dissemination of such information.

Certainly, publishing information about received requests could be a security issue in its own right. Fortunately, it is not necessary to disclose full requests' URLs to other peers. Since each host needs just to check if the URL it marked as suspicious has also been reported by other hosts, it is enough if the hosts compare the *hashes* (digests) of the requests' URLs. Hashes (like SHA1) are generated using one-way functions and are generally deemed irreversible. Even though findings are occasionally published about the weaknesses of certain hashing methods [20], there are always algorithms available that are considered safe, even for the storage of critical data (at the time of writing two examples are SHA-2 and MD6). If a host only publishes hashes of the requests that it deems suspicious, it allows other hosts to verify their findings, but at the same time protects its potentially confidential information.

For the proposed method to work, it is enough if hosts just publish a list of hashes of the received requests that they consider suspicious, but other data may improve the interoperability and possible future heuristics. Therefore, it is suggested that hosts use a structured document format, e.g. based on JSON [10]. A sample published document may look like the one in Listing 1.1. This will be referred to as a *List of Suspicious Requests*, abbreviated *LSR*. The list presented in Listing 1.1 is a part of a list taken from an actual running instance of the reference implementation (and includes the mentioned additional information).

The explanation of all the elements in this LSR is provided below:

- (C) denotes the class of the information. This can either be Original or Forward from another peer.
- (A) is the age of the suspicious request, e.g. how many hours ago it was received.
- (MD5) indicates the algorithm used and is followed by the resultant hash of the request.
- (R) includes the actual request received and is presented here for informational/debug purposes only.

Listing 1.1. Sample LSR with additional debug information

```
{ C:0, A:57, MD5:2cf1d3c7fe2eadb66fb2ba6ad5864326, R:"/pacpdvlgj.html" }
{ C:0, A:53, MD5:2370f28edae0afcd8d3b8ce1d671a8ac, R:"/statsa/" }
{ C:F, A:32, MD5:2f42d9e09e724f40cdf28094d7beae0a }
{ C:F, A:31, MD5:8f86175acde590bf811541173125de71 }
{ C:F, A:24, MD5:eee5cd6e33d7d3deaf52cadeb590e642 }
{ C:0, A:17, MD5:bd9cdbfedca98427c80a41766f5a3783, R:"/Docs/ads3.html" }
```

2.3   Maintenance of the Lists

For the process to work as intended, each server must not only identify suspicious requests, but also generate and publish the list of them and retrieve similar lists from other servers. However, exchanging of the LRS leads to an issue of data retention, and two questions need to be answered:

- How long should an LSR contain an entry about a suspicious request after such a request was received.
- Should the hashes received from other peers be preserved locally if the originating server does not list them anymore, and if so, for how long?

Both issues are related to the load (i.e. number of requests per second) received either by the local or the remote server. Servers with very high loads and a high number of suspicious requests will likely prefer shorter retention times, while niche servers may only have a few suspicious requests per week and would prefer a longer retention period. The results of experiments presented later in this article illustrate how the retention period may impact the quality of detection, and more results may be obtained if the method becomes more widely adopted.

3   Implementation and Test Environment

A reference implementation has been prepared to verify the feasibility of the proposed method and to prove this concept. The application has been programmed in Java and have been tested using the data from a few real web servers. The architecture of the application is presented in Fig. 3.

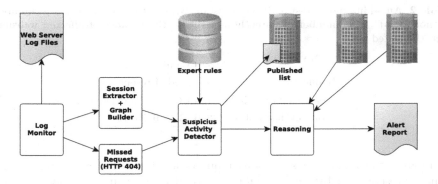

Fig. 3. The architecture of the reference implementation

The requests for unavailable resources (e.g. these that result in HTTP 4xx response) are directly forwarded to the aggregation of suspicious requests. The original application also includes a behavior-based anomaly detection module, which is outside the scope of this article, but has been described in [23].

The *Suspicious Activity Detector* module aggregates all suspicious requests and matches them against some pattern-based rules, to eliminate requests that are well known and harmless but are often missing on servers. The list of requests, together with their occurrence frequency is presented in Table 1. They can safely be ignored and not reported to the administrator. This process is referred to as "white-listing".

In addition to this general white-listing, applicable for all the web sites, a website may benefit from additional white-listing of specific pages, which could generate false-positives. During the tests, it was found out that three of the tested websites had a relatively large number of reports for just two URLs (see Table 2). It is likely that the websites' structure might had changed during the testing

Table 1. A white-list of requests that are considered legitimate, yet often result in a "not found" (HTTP 4xx) response.

Request	Total number		Comment
	Not found	Present	
/ & index.html	1,342	161,542	Home page of a website
favicon.ico	24,899	4,067	The website's icon
apple-touch-icon*.png	2,255	112	Icons used by IOS-based devices
robots.txt	14,728	16,712	*robots.txt* and *sitemap.xml* are
sitemap.xml	111	493	file looked for by search engines

period as both web-pages are often present in commercial websites. Such change would also explain the number of request attempts for these missing resources.

Table 2. An additional, site-specific white-list of requests that generate a large number of "not found". The number of parenthesis indicate the number of affected websites from the tested set.

Request	Total number	
	Not found	Present
/services.[html|php]	153 (3)	4,219 (1)
/contact.[html|php]	1,230 (3)	9,810 (2)

Finally, the remaining suspicious requests are formatted according to the sample presented in Listing 1.1, and are stored in a document inside the web server directory to be accessible by other hosts.

3.1 Reasoning

The last module (named 'Reasoning') receives the current request (if it is considered suspicious) and, at the same time, periodically retrieves lists of such suspicious requests from other hosts. With each request, the module has to decide whether it should report it for human (e.g. administrator's) attention or not. There are various strategies that can be implemented here, depending on the type of application being protected and the number and the type of the peer servers it receives the data from. Some basic strategies are discussed in the next chapter, yet the system administrator may be willing to adjust the system's response depending on their own requirements.

4   Test Method and Achieved Results

For the purpose of evaluating the proposed algorithm, a number of log files have been acquired, as listed in Table 3. These log files came from a number of

Table 3. Number of lines per log file used for the test experiments. Only parseable, validated lines were counted. Bottom row lists the number of requests that resulted in error response 4xx or 5xx.

	Site 1	Site 2	Site 3	Site 4	Site 5	Site 6	Site 7
All requests	311,530	1,030,186	108,859	53,271	418,361	254,638	886,233
4xx and 5xx	14,861	287,394	4,017	14,281	41,706	7,381	25,885

unrelated servers, located in several countries. Due to the nature of the method, the logs must cover the same period of time, in order to simulate a specific real time period. All these logs cover the same period of approximately 1 year.

Based on these logs, a simulation environment was prepared, that emulated these web servers over the specific time period. Each emulated server analyzed its log files and published the LSR for the other servers to download. To clearly identify the benefits of the collective detection, and eliminate the impact of other methods, all and only requests that resulted in HTTP response 4xx ("not available") or 5xx ("server error") were listed in the server's LSR. The overall number of such requests is presented in Table 3. Additionally, the changes in number of these requests over consecutive weeks is illustrated in Fig. 4. As can be seen there, these numbers can vary from week to week even by an order of a magnitude.

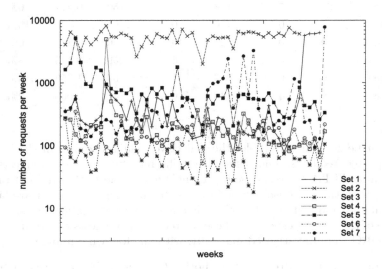

Fig. 4. Change in number of requests resulting in 4xx or 5xx error response over time. Aggregated per week for each of the server log sets.

The results achieved with the collective detection were also compared with the results of an analysis performed by LORG [13], an Open-Source project designed to identify malicious requests in a web server's log files.

Several scenarios were tested, with different configuration, in order to determine their impact on the quality of detection. These include evaluating the impact of the number of web servers participating in the exchange and changing how long the suspicious requests are kept in the LSRs.

4.1   The Baseline Scenario

In this scenario, all seven servers were emulated, and the LSRs kept records of suspicious requests for 7 days. The results of the collective detection for each server is presented in Table 4, together with the result of the LORG analysis.

Table 4. Results of the baseline scenario, number of reported suspicious requests by LORG and collective detection, split into these which ended up as "unavail." (4xx and 5xx response) and these that resulted in 2xx response ("present"). The number of reported incidents is also presented as a per mille (‰) of total log entries.

Response:	LORG & CD		LORG only		Collective detection		
	Present	Unavail.	Present	Unavail.	Present	Unavail.	Ratio
Site 1	0	3	0	0	10	1,030	3‰
Site 2	23	5	148	8	150	6,411	6‰
Site 3	0	0	0	0	1,430	165	15‰
Site 4	0	0	28	16	0	306	6‰
Site 5	0	0	19	3	259	3,143	8‰
Site 6	0	0	7	1	63	1,356	6‰
Site 7	0	4	43	0	47	2,000	2‰

As can be seen in this table, the collective analysis results in significantly more reported requests. The results are split into two groups - requests that resulted in a HTTP 2xx response ("present") and requests that did not process correctly, with the response 4xx or 5xx sent to the client. The later are labeled as 'unavailable' and can be safely classified as scans or attack attempts.

Some of the results, especially these related to the requests reported as harmful by LORG were investigated manually, with the following findings:

- Most of the requests reported by both LORG and the collective detection were requests for a home page, but with additional parameters intended to alter its operation; like /?include=../../../etc/passwd.
- Majority of the requests reported by LORG but not the collective detection were actually false positives. This includes almost all of the 148 requests resulting in "present" response from Site 2.

– The unusual number of "present" responses for reported requests in Site 3 is a result of the configuration of this website, which returned a page even for unexpected URLs, unless they were severely malformed.

Most common examples of requests reported by the collective detection and LORG are presented in Table 8. Since this table presents the most popular examples (every one occured several hundred times), the requests may appear quite ordinary. Collective detection is however able to also identify more sophisticated attacks, as illustrated in Table 5.

Table 5. Less common examples of malicious requests identified by collective detection.

Example URL
/includes/fckeditor/editor/filemanager/upload/php/upload.php
/includes/uploadify/uploadify.swf
/index.php?m=admin&c=index&a=login&pc_hash=
/wp-content/themes/felis/download.php?file=../../../wp-config.php
/?page_id=\"><script> alert(\"m3t4l&master\");</script>
/go?to=http://pastebin.com/raw.php?i=pA3y1PSN
/cgi-bin/php-cgi?%2D%64+%61 %6C%6C%6F%77 %5F%75 %72 %6C%5F%69 (...)
/asp/freeupload/uploadTester.asp
/beheer/editor/assetmanager/assetmanager.asp

Statistical analysis of the requests reported by the collective detection resulted in identification of the most common attack vectors, which are presented in Table 6.

Table 6. Most common attack vectors identified by collective detection.

Occurences	Description
10,332	Attempts to locate or use basic WordPress URLs
2,049	Requests for various /admin/ URLs (excluding WordPress)
1,679	Attempts to locate CKEditor
265	Attempts to identify or exploit WordPress xml-rpc
163	Attempts to locate OpenFlashChart
150	Direct attempt to manipulate parameters of the primary web page; e.g. /?(...)params or /index.php?(...)params
107	Requests targeting Uploadify
43	Attempts to access PhpMyAdmin

4.2   Impact of the LSR History Length and the Number of Peers

It is rather obvious, that the effectiveness of the detection depends on the number of peers involved in the exchange and how long they keep the history of the suspicious requests. Table 7 presents the results of the collective assessment for smaller group of peers. Interestingly, the overall traffic processed by the participating websites has much more impact on the final results than the number of peers involved.

Table 7. Decrease in number of reported requests against the baseline scenario (percent of the baseline reported request) for decreased number of peers participating in the collective assessment.

	6 peers	5 peers	4 peers	3 peers	2 peers
High-traffic sites	95.8 %	94.3 %	85.8 %	86.7 %	61.1 %
Low-traffic sites	91.7 %	77.5 %	56.3 %	37.9 %	21.8 %

The impact of the history length is presented in Fig. 5. As can be seen there, there is very little gain in number of reported requests for history length longer than 10 days. It seems that 5–10 days is the maximum a host would need. High-traffic servers may need to settle for a 1–2 day history, but that still provides substantial benefits for the collective detection.

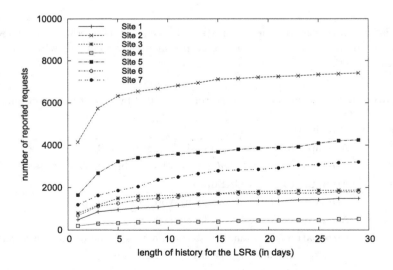

Fig. 5. Number of suspicious request reported by each site as the function of LSR history length.

Table 8. Sample most common findings of potentially malicious requests by LORG and collective detection.

Example URL
Detected by the collective assessment, but not by LORG
/administrator/index.php
/admin.php
/?q=user/register
/wp-admin/admin-ajax.php
/wp-login.php
/wp-login.php?action=register
/xmlrpc.php
Detected by LORG, but not the collective assessment
/pl/?page=http://www.lincos.eu/cache/mod_custom/ID-RFI.txt????
/wordpress/wp-admin/load-scripts.php?c=1&load%5B%5D=hoverIntent,common, admin-bar,svg-painter&ver=4.2.2
/?x=()
Detected by both LORG and the collective assessment
/wp-admin/admin-ajax.php?action=revslider_show_image&img=../wp-config.php
/?x=() { :; }; echo Content-type:text/plain;echo;echo;echo M'expr 1330 + 7'H;/bin/uname -a;echo @
/?page_id=../../../../../../../../../../etc/passwd
/upload.asp?action=save&type=IMAGE&style=standard'%20and%201=2 %20 union%20select%20S_ID,S_Name,S_Dir,(...)

5   Conclusions and Future Work

The method proposed in this paper aims at providing automated identification of potentially harmful requests with the minimum level of involvement from the system administrators. Suspicious requests are presented to other peers; i.e. web servers, which participate in information exchange. A report of a suspicious activity is produced when identical requests had been encountered and identified as suspicious by other peers.

The method may significantly hinder the *modus operandi* of most malware, due to the fact that after the few initial attempts to attack a number of servers, it will become increasingly difficult for a malware to successfully attack new ones. The servers will recognize the attacks because of the knowledge acquired from their peers. This way, the whole system will develop an immune response similar to the one observed in living organisms.

Emulation-based evaluation of the method, utilizing real data acquired from seven web servers showed that the median ratio of reports were 0,6 % of all the requests. This translates to an average of 5–10 suspicious requests per day brought to the administrators' attention, which is an acceptable level, making analysis of the log files feasible.

Analysis of the data also showed, that most of these request can be grouped into a few common attack vectors. For example, attempts to locate WordPress specific web pages accounted for almost $2/3$ of all the reported attempts. Another 10 % were attempts to locate CKEditor - a commonly used module, but having numerous vulnerabilities. An administrator can thus easily reduce the number of reported attempts, by an order of magnitude, if they know they do not have the specific module, nor care about such probing attempts.

Achieved results were compared to LORG, an open source software designed to identify attack attempts by analyzing the URLs of the incoming requests. Surprisingly, a major part of LORG's findings were identified as false positives. It was due to the fact, that two of the websites used large and complicated forms that resulted in a long array of parameters passed with the requests, which were considered suspicious by LORG. LORG also did not identify any scans/probes that did not have an arguments sent along with the URL. This is due to the nature of its analysis, yet they constituted the majority of the identified attempts.

In terms of the computational power required for the described method, the reference single-thread implementation (in Java) was able to process over 30,000 requests (log entries) per second on a typical PC (Intel, 4 GHz). It shall therefore not add a significant workload for the web server.

The method described in this paper can provide substantial benefits to the security of websites right now. It also opens new paths of research that could lead to its further refinement or specialized applications. The development of the decision algorithm will presumably provide the most benefits for the system, since improvement to the local reasoning and identification of the suspicious request will reduce the amount of data exchanged between the peers and will speed up each assessment. By introducing local user tracking and their behavior analysis, the system could also be able to distinguish "blind" scans (which are usually less harmful) from targeted attack attempts. This may be a benefit for high-traffic sites, however will likely result in a delayed detection of attack attempts for the group of servers participating in the data exchange network.

References

1. Agosti, M., Crivellari, F., Di Nunzio, G.: Web log analysis: a review of a decade of studies about information acquisition, inspection and interpretation of user interaction. Data Min. Knowl. Disc. **24**(3), 663–696 (2012)
2. Anderson, R., Barton, C., Böhme, R., Clayton, R., Van Eeten, M.J., Levi, M., Moore, T., Savage, S.: Measuring the cost of cybercrime. In: Böhme, R. (ed.) The Economics of Information Security and Privacy, pp. 265–300. Springer, Heidelberg (2013)
3. Auxilia, M., Tamilselvan, D.: Anomaly detection using negative security model in web application. In: 2010 International Conference on Computer Information Systems and Industrial Management Applications (CISIM), pp. 481–486 (2010)

4. Florêncio, D., Herley, C.: Sex, lies and cyber-crime surveys. In: Schneier, B. (ed.) Economics of Information Security and Privacy III, pp. 35–53. Springer, Heidelberg (2013)
5. García-Teodoro, P., Díaz-Verdejo, J., Maciá-Fernández, G., Vázquez, E.: Anomaly-based network intrusion detection: techniques, systems and challenges. Comput. Secur. **28**(1–2), 18–28 (2009)
6. van Goethem, T., Chen, P., Nikiforakis, N., Desmet, L., Joosen, W.: Large-scale security analysis of the web: challenges and findings. In: Holz, T., Ioannidis, S. (eds.) Trust 2014. LNCS, vol. 8564, pp. 110–126. Springer, Heidelberg (2014)
7. Han, E.E.: Detection of web application attacks with request length module and regex pattern analysis. In: Genetic and Evolutionary Computing: Proceedings of the Ninth International Conference on Genetic and Evolutionary Computing, 26–28 August 2015, Yangon, Myanmar, vol. 2, pp. 157. Springer, Switzerland (2015)
8. Hyman, P.: Cybercrime: it's serious, but exactly how serious? Commun. ACM **56**(3), 18–20 (2013)
9. Iváncsy, R., Vajk, I.: Frequent pattern mining in web log data. Acta Polytechnica Hungarica **3**(1), 77–90 (2006)
10. JSON: a lightweight data-interchange format. http://www.json.org
11. Kruegel, C., Vigna, G., Robertson, W.: A multi-model approach to the detection of web-based attacks. Comput. Netw. **48**(5), 717–738 (2005)
12. McAfee: Net Losses: Estimating the Global Cost of Cybercrime (2014). http://www.mcafee.com/us/resources/reports/rp-economic-impact-cybercrime2.pdf
13. Muller, J.: Implementation of a Framework for Advanced HTTPD Logfile Security Analysis, Master's thesis (2012)
14. Netcraft: Web Server Survey (2015). http://news.netcraft.com/archives/2013/11/01/november-2013-web-server-survey.html
15. OWASP: Web Application Firewall. https://www.owasp.org/index.php/Web_Application_Firewall
16. Pałka, D., Zachara, M.: Learning web application firewall - benefits and caveats. In: Tjoa, A.M., Quirchmayr, G., You, I., Xu, L. (eds.) ARES 2011. LNCS, vol. 6908, pp. 295–308. Springer, Heidelberg (2011)
17. Rieck, K., Laskov, P.: Language models for detection of unknown attacks in network traffic. J. Comput. Virol. **2**(4), 243 (2007)
18. Roesch, M.: Snort: lightweight intrusion detection for networks. In: LISA, USENIX, pp. 229–238 (1999)
19. Salama, S.E., Marie, M.I., El-Fangary, L.M., Helmy, Y.K.: Web server logs preprocessing for web intrusion detection. Comput. Inf. Sci. **4**(4), p123 (2011)
20. Stevens, M.: Advances in hash function cryptanalysis. ERCIM News **2012**(90), 26–27 (2012)
21. Symantec: Internet Security Threat Report (2015). http://www.symantec.com/security_response/publications/threatreport.jsp
22. WhiteHat: Website Security Statistics Report (2013). http://info.whitehatsec.com/2013-website-security-report.html
23. Zachara, M.: Collective detection of potentially harmful requests directed at web sites. In: Hwang, D., Jung, J.J., Nguyen, N.-T. (eds.) ICCCI 2014. LNCS, vol. 8733, pp. 384–393. Springer, Heidelberg (2014)

A Grey Approach to Online Social Networks Analysis

Camelia Delcea[1(✉)], Liviu-Adrian Cotfas[1], Ramona Paun[2],
Virginia Maracine[1], and Emil Scarlat[1]

[1] The Bucharest Academy of Economic Studies, Bucharest, Romania
{camelia.delcea, emil_scarlat}@csie.ase.ro,
liviu.cotfas@ase.ro, virginia.maracine@casie.ase.ro
[2] Webster University, Bangkok, Thailand
paunr@webster.ac.th

Abstract. Facebook is one of the largest socializing networks nowadays, gathering among its users a whole array of persons from all over the world, with a diversified background, culture, opinions, age and so on. Here is the meeting point for friends (both real and virtual), acquaintances, colleagues, team-mates, class-mates, co-workers, etc. Also, Facebook is the land where the information is spreading so fast and where you can easily exchange your opinions, feelings, travelling information, ideas, etc. But what happens when one is reading the news feed or is seeing his Facebook friends' photos? Is he thrilled, excited? Is he feeling that the life is good? Or contrary: he is feeling lonely, isolated? Is he doing a comparison with his friends? These are some of the questions this paper in trying to answer. For shaping some of these relationships, the grey system theory will be used.

Keywords: Grey incidence analysis · Social networks · Facebook · Correlation analysis

1 Introduction

Since the appearance of the first online social network (OSN), in 1997, Six-Degrees, a large number of other social networks such as Twitter, Facebook, LinkedIn, QZone, Google+, Instagram etc. have become popular platforms where more than one and a half billion people are gathering and connecting [1]. It is estimated that, in the future, their number will continue to increase, reaching a total of 2.44 billion users in 2018 [2].

Facebook is one of the largest socializing networks nowadays, gathering among his users a more than 864 million daily active users from all over the world, with a diversified background, culture, opinions, age and so on. Facebook's mission: both "to give people the power to share and make the world more open and connected" and to enable people "to stay connected with friends and family, to discover what's going on in the world, and to share and express what matters to them" [3]. Here is the meeting point for friends (both real and virtual), acquaintances, colleagues, team-mates, class-mates, co-workers, etc. Also, here is the land where the information is spreading so fast and where one can easily exchange its opinions, feelings, traveling information,

© Springer-Verlag Berlin Heidelberg 2016
N.T. Nguyen and R. Kowalczyk (Eds.): TCCI XXII, LNCS 9655, pp. 60–79, 2016.
DOI: 10.1007/978-3-662-49619-0_4

ideas, etc. That is just one of the reasons why people are becoming more and more attached to this social network.

But, as this network implies first of all the people, the way all this information is transferred from a person to another is very different. Sometimes one can enjoy and be happy for someone else's success, but, in the same time it can be annoyed by another persons' activities, social life, professional life, etc.

This paper tries to see whether the people from a randomly chosen sample are comparing themselves with the ones in their own network by considering the posts their friends are making on Facebook (including here the information posted on news feed, their photos, etc.) and whether there is an incidence between the social comparison orientation and the appearance of a negative feeling about themselves. Moreover, the connection between the number of Facebook friends and the frequency of using this social network is analysed. For this, the correlation analysis will be used along with the grey theory incidence analysis.

2 Now-a-Days Social Networks Analysis

One of the phenomena encountered in the now-a-days reality in social networks is the spreading speed of any type of information within the social network and can be very good explained through the so-called "go viral" property [4]. Accordingly to this property, not only that the information flow has an enormously highly increased speed, but also, the services broadcasted through the social networks are getting very fast to their end user.

As becoming part of our every-day-life, the social networks have an important impact on our behaviour, thoughts, ways of action, state of being, etc.

For this reason, the studies on different aspects related to the social networks have increased recently, some of them referring the technical aspects, such as:

- Networks' stability [4];
- Anomaly detection [5];
- Creating overlapping community identification algorithms [6–8];
- Competitive contagion and adoption dynamics [9];
- Scalable secure computing in these networks [10];
- Recommender systems based on social networks [11]

while others are focusing on practical and social aspects:

- Social comparison [12];
- Social capital [13];
- Social well-being [14];
- Social activity [15];
- Social structure [16, 17];
- Personality and Facebook use [18, 19];
- Self-presentation in social networks [20];
- Decision-making [21];
- Segregation within these networks [22], etc.

This study continues the idea of social comparison existence in Facebook relationships presented by Lee in his work [12] and replicate its study in order to see whether this relationship still exists even on another continent on the persons which have similar characteristics with the ones considered, such as age, background, interests, etc. For this, the results obtained through the questionnaire are compared to the ones of Lee. More, a grey incidence analysis on the considered variables has been applied in order to sustain, even more, the obtained results.

3   Grey Knowledge in Online Social Networks

Starting from the idea that in the online social environments the human component becomes the central actor, playing a tremendous role in the achieving, processing and transferring knowledge in such a network, a new type of knowledge is developing here, namely, the grey knowledge [23].

This knowledge is situated at the border of the two classical types of knowledge: the tacit and the explicit one, having a dynamic evolution, fuelling the ties created in both online and offline networks. More, the grey knowledge is characterizing the feedback loops created in the online environment through the messages, status, pictures or feelings expressed here. Even more, Heidemann et al. (2012) concluded that the information shared through the OSN is a conscious act, people deliberately deciding what information to share in term of importance for its own image across the network or for the people that might be interested in [1].

Therefore, with this new type of knowledge a shift has been produced in the knowledge society by passing from the implicit knowledge represented by the personal knowledge (intention, heuristics, rules of thumb, personal skills, know-how, etc.) and organizational knowledge (routines, culture, history, shared models, stories, ways of thinking, etc.) to the grey knowledge.

In the same time, the grey knowledge can be found at the edge of the explicit knowledge represented by both the organizational artefacts (designs, reports, handbooks, manuals, tutorials, etc.) and the collective knowledge (attitudes, comportments, rules, norms, practices, habits, inter-relational communications, etc.). Therefore, it can be concluded that this new type of knowledge can be associated with the internal processes (experience, reflection, evaluation, observing, intuition, emotion, application of talents, etc.) or with the external ones (synchronous discussion, chatting in social platforms, contacting users, sending e-mails, team interaction, etc.).

Even more, the grey knowledge is present in the online activities, in argumentations and comments made on product sales websites, on blogs, in the public or private messages made by the online social networks users, in the Twitter's tweets, etc.

Individuals, the main component of the online social networks, are very distinct one of another and over the time they tend to be unpredictable, have a personal way to respond to external stimuli, have their own opinion regarding a specific situation, are unique, are capable of innovation, all of these being the result of the free-will, self-awareness, conscience, imagination [24, 25].

More, by randomly selecting some of the Facebook users' social networks, it can easily be observed that these networks come in so many different shapes, as they can be considered a personal online "finger-print". Figure 1 is presenting some examples of personal Facebook networks extracted using Gephi 0.8.2.

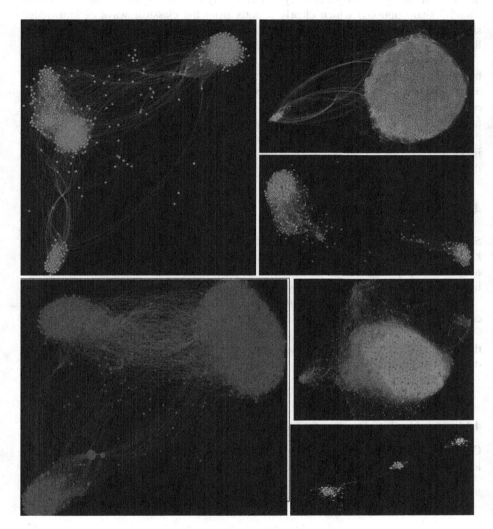

Fig. 1. Examples of personal Facebook networks

The now-a-days online social network reality proves that the grey knowledge that is passing through the network's ties can be very well explained through the "go viral" property [4]. This property is shortly explaining the extremely high speed of the information which is passing through the social networks and how it succeed in a relatively short period of time to come across a large number of network's members.

Moreover, the interpersonal sharing character of the users, which are consciously deciding what information to share in the online communities and to whom it will be given [26] is shaping even more the notion of grey knowledge.

Now-a-days, the social platforms like Facebook, for example, allow people to share this kind of information both directly and indirectly (see the "friends" and "friends-of-friends" share options) which clearly departs from the classical ways of knowledge sharing. Even more, the grey knowledge is a function of social contextual properties like relationship qualities and position within a network [27, 28].

In this whole context, the consumers' decision and perceptions become influenced by the other online social networks members' opinions, experiences and thoughts about a certain company, brand or product. For this, the present paper tries to shape the relationship between how a consumers is taking his consuming decisions in a non-online social environment and how its perception about a product is influenced by the other members he interacts with in online environments. On this purpose, a grey incidence analysis is conducted, knowing that the grey systems theory works better in this kind of environments, with a high degree of uncertainty.

4 Grey Incidence Analysis

Starting from the grey systems general definition as being a mix of information, partly known and partly unknown, it can easily be transferred this grey property to the social networks, especially because in these kind of networks the main component is the human one, greatly characterized by uncertainty in behaviour and decisions.

Grey incidence analysis is a central piece of grey system theory and it also can be considered the foundation for grey modelling, decision making and control [29].

Over time, a permanent interest manifested on this method led researchers from different parts of the world to study and extended it, which has conducted to the development of different other types of grey incidence [30].

The classical degrees of grey incidence are computed as in the following.

4.1   The Absolute Degree of Grey Incidence

Considering two sequences of data with non-zero initial values and with the same length, data X_0 and X_j, j = 1...n, with t = time period and n = variables: [30]

$$X_0 = \left(x_{1,0}, x_{2,0}, x_{3,0}, x_{4,0}, \ldots, x_{t,0}\right), \tag{1}$$

$$X_j = \left(x_{1,j}, x_{2,j}, x_{3,j}, x_{4,j}, \ldots, x_{t,j}\right), \tag{2}$$

The zero-start points' images are:

$$X_j^0 = \left(x_{1,j} - x_{1,j}, x_{2,j} - x_{1,j}, \ldots, x_{t,j} - x_{1,j}\right) = \left(x_{1,j}^0, x_{2,j}^0, \ldots, x_{t,j}^0\right) \tag{3}$$

The absolute degree of grey incidence is:

$$\varepsilon_{0j} = \frac{1 + |s_0| + |s_j|}{1 + |s_0| + |s_j| + |s_0 - s_j|} \tag{4}$$

with $|s_0|$ and $|s_j|$ computed as follows:

$$|s_0| = \left| \sum_{k=2}^{t-1} x_{k,0}^0 + \frac{1}{2} x_{t,0}^0 \right| \tag{5}$$

$$|s_j| = \left| \sum_{k=2}^{t-1} x_{k,j}^0 + \frac{1}{2} x_{t,j}^0 \right| \tag{6}$$

4.2   The Relative Degree of Grey Incidence

Having two sequences of data with non-zero initial values and with the same length, X_0 and X_j, $j = 1 \ldots n$, with t = time period and n = variables: [30]

$$X_0 = (x_{1,0}, x_{2,0}, x_{3,0}, x_{4,0}, \ldots, x_{t,0}), \tag{7}$$

$$X_j = (x_{1,j}, x_{2,j}, x_{3,j}, x_{4,j}, \ldots, x_{t,j}), \tag{8}$$

The initial values images of X_0 and X_j are:

$$X_0' = \left(x_{1,0}', x_{2,0}', \ldots, x_{t,0}' \right) = \left(\frac{x_{1,0}}{x_{1,0}}, \frac{x_{2,0}}{x_{1,0}}, \ldots, \frac{x_{t,0}}{x_{1,0}} \right) \tag{9}$$

$$X_j' = \left(x_{1,j}', x_{2,j}', \ldots, x_{t,j}' \right) = \left(\frac{x_{1,j}}{x_{1,j}}, \frac{x_{2,j}}{x_{1,j}}, \ldots, \frac{x_{t,j}}{x_{1,j}} \right) \tag{10}$$

The zero-start points' images calculated based on (9) and (10) for X_0 and X_j are:

$$X_0^{0'} = \left(x_{1,0}' - x_{1,0}', x_{2,0}' - x_{1,0}', \ldots, x_{t,0}' - x_{1,0}' \right) = \left(x_{1,0}'^0, x_{2,0}'^0, \ldots, x_{t,0}'^0 \right) \tag{11}$$

$$X_j^{0'} = \left(x_{1,j}' - x_{1,j}', x_{2,j}' - x_{1,j}', \ldots, x_{t,j}' - x_{1,j}' \right) = \left(x_{1,j}'^0, x_{2,j}'^0, \ldots, x_{t,j}'^0 \right) \tag{12}$$

The relative degree of grey incidence is computed as:

$$r_{0j} = \frac{1 + |s_0'| + |s_j'|}{1 + |s_0'| + |s_j'| + |s_0' - s_j'|} \tag{13}$$

with $|s'_0|$ and $|s'_j|$:

$$|s'_0| = \left| \sum_{k=2}^{t-1} x'^0_{k,0} + \frac{1}{2} x'^0_{t,0} \right| \tag{14}$$

$$|s'_j| = \left| \sum_{k=2}^{t-1} x'^0_{k,j} + \frac{1}{2} x'^0_{t,j} \right| \tag{15}$$

4.3 The Synthetic Degree of Grey Incidence

The synthetic degree of grey incidence is based on both the absolute and the relative degrees of grey incidence: [30]

$$\rho_{0j} = \theta\varepsilon_{0j} + (1 - \theta)r_{0j}, \tag{16}$$

with j = 2,..., n, $\theta \in [0, 1]$ and $0 < \rho_{0j} \leq 1$.

With these, the grey incidence will be applied in the next section to the data gathered through a questionnaire regarding the Facebook activity.

5 Case Study

Starting from a recent case study conducted by Lee [12] on how the people are comparing themselves with others on social network sites, with application on the Facebook network, this paper is redoing the same analysis in similar conditions. The purpose of this study is to see whether the results obtained in [12] can be generally valid within any Facebook community which has almost the same characteristics.

5.1 The Questionnaire

Even though the number of questions was quite large, in this paper it is only presented the set of questions that are similar with the one used in the mentioned study. The types of questions was mixed: there have been both open and closed questions, multiple choice questions and yes-no questions. A 5-point Likert scale was used to evaluate the answers received, ranged from 1 (strongly disagree) to 5 (strongly agree).

Some of these questions were:

- Personal data:

    - Age;
    - Sex;
    - Study year.

- Split question: Do you use Facebook? – for a "no" answer, the questionnaire was over, while for an "yes" answer it continues with the following questions:
- Number of Facebook friends:
  - How many friends do you actually have on Facebook?
  - Approximately, how many among these were/are your college colleagues?
  - How many of your Facebook friends are also friends with you in the real life?
  - With how many among all your Facebook friends you are usually communicating frequently? (at school, on Facebook, in your spare time, etc.)
  - How many of them, do you consider to be your close friends?
  - How often do you communicate with your close friends?
  - How often do you communicate face-to-face with your close friends?
- Social comparison on Facebook (expressed through frequency) and in real life:
  - I often compare myself with my other Facebook friends while I am reading news feed.
  - I often compare myself with my other Facebook friends while I am viewing their photos and visited places.
  - I usually make comparisons between my dearest ones and the other persons in my group of friends.
  - I observe my behaviour in different situations and I often compare it with others' behaviour in similar situations.
- Self-esteem, uncertainty, anxiety and believes:
  - I think I am worthy person, at least as my friends.
  - I think I have plenty of qualities.
  - I think others are usually appreciating me for my work.
  - I think others are trusting my decisions and ideas.
  - In general, my opinions about myself are different from the others.
  - My opinion about myself is different from one day to another.
  - I usually change the opinion about myself during a day, depending on the encountered events.
  - Unpredictable events are irritating me.
  - I feel frustrated when I have lack of information.
  - I get nervous quite easily.
  - When dealing with unexpected situations, I become angry and irritated.
  - Last week, things that usually are not irritating me, bothered me.
  - Lately, I felt extremely unhappy, even though the dearest ones tried their best to get me out of this situation.
- Depression and negative feelings:
  - Sometimes, while reading the news feed on Facebook, I think the others have a better life than me.
  - Sometimes, while watching my friends' photos on Facebook, I think the others have a better life than me, that they are more happily and are more enjoying their lives.
  - Sometimes, while reading the postings on Facebook, I think the others are doing so much better than I do.

- Sometimes, while reading the posting on Facebook, I feel lonely and isolated from the world.
- Facebook use and expectations:
  - Facebook is a part of my daily routine.
  - I feel "out of reality" when I stay away from Facebook for a period of time.
  - Usually, I connect on Facebook:
    - I stay connected all day long;
    - A couple of times a day;
    - Once a day;
    - Once a week;
    - Once at every few weeks.
  - When I post something, I expect that the others will positively respond to it.
  - If none of my friends reacts to my posting, I feel sad.

5.2   Results from Romania

For this, in period 1st–15th March 2014, the students of The Bucharest University of Economic Studies, Faculty of Economic Cybernetics, Statistics and Informatics, have voluntary participated on this survey and they were ask to sincerely answer to a series of questions presented below, most of this questions being similar to the ones used in Lee's study [32].

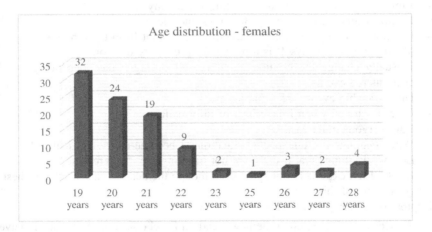

Fig. 2. Female's age distribution

The number of respondents was 144, with an age distribution range between 19 and 28 years old, 33.33 % of them being male and 66.67 % of them being females. Their distribution on study year is: 57 % first year, 16 % second year and 27 % third year, while their gender distribution can be seen in Figs. 2 and 3.

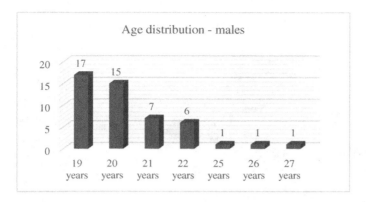

Fig. 3. Male's age distribution

By processing the personal data, the following results have been gathered: the average age of the sample is 20.7 years old, and only 1 person from 144 is not using Facebook, representing less than 1 % of the whole sample, reducing the sample to 143 valid respondents.

Moreover, by analysing the answers gathered on the questionnaire, it seems that a person has, on average, almost 671 friends on Facebook, 118 among them being college colleagues. Also, on average, the respondents have said that approximately 134 of the Facebook friends are persons they have met in real life. The respondents also pointed that only with 52 of these friends they succeed to communicate frequently (at school, on Facebook, etc.). Even more, the respondents, are considering, that on average, only 12 of these friends are close friends, their range being between 1 and 150.

As for the communication with close friends using Facebook, the respondents have said that, on average, they are communicating quite often with these ones: 3.75 points on a 5 point Likert scale, while the face-to-face communication with the close friends is a more frequent communication way, reaching, in medium, 4.19 points from a 5-point Likert scale.

5.3   Results from Thailand

The study in the Webster University, Bangkok, Thailand was conducted in 15th–30th November 2014 and it kept the age-gender structure from the previous study on Romania. Therefore, the 143 participants on the study were between 19–28 years old and all of them said that they have a Facebook account. The distribution on study year is represented in Fig. 4.

After processing the personal data, it has been found that, on average, a person has 602 friends on Facebook, 109 among them being college colleagues and approximately 339 of the Facebook friends are persons they have met in real life. Also they pointed that only with 53 of these friends they succeed to communicate frequently (at school, on Facebook, etc.) and only 15 of these friends, on average, are close friends, their range being between 0 and 78.

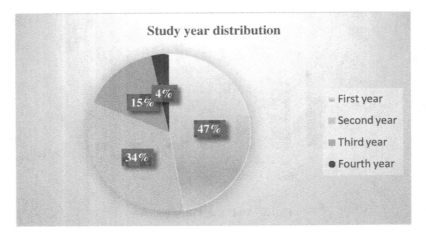

Fig. 4. Study year distribution - Thailand

More, regarding the communication with close friends using Facebook, the respondents have said that, on average, they are communicating quite often with these ones: 3.66 points on a 5 point Likert scale, while the face-to-face communication with the close friends is a less frequent communication way, reaching, on average, 3 points from a 5-point Likert scale.

By making a comparison between the students from the two considered countries, it can be seen that the questionnaire responds are quite similar, with few exceptions (see Fig. 5).

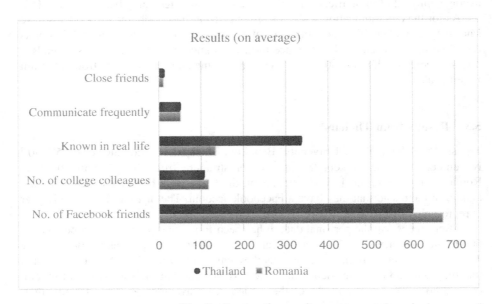

Fig. 5. Comparative results

For example, the number of close friends, the number of the friends to whom they are frequently communicating, the number of college colleagues and, to a certain point, even the number of friends, on average, are comparable for the two countries.

One small difference can be noticed here related to the number of the Facebook fiends the users are actually knowing from the real life. In this case it can be seen that only 134 real life friends were registered for the Romanian students, while for the Thai students the number is significantly much bigger: 339 (Fig. 5). This aspect may be due to the cultural differences, way of life, openness, etc. and will necessitate a further investigation.

As for the communication with their friends, it can be concluded that the frequency of Facebook communication is almost the same: 3.75 versus 3.66 (see Fig. 6) on a 5-point Likert scale, indicating a considerable degree of communication made through this network.

Another difference is encountered related to the face-to-face conversation, which is more frequently in the case of the Romanian students, reaching 4.19 points on a 5-point Likert scale, while for the Thai students is just an occasionally state, getting only 3 of the 5 points (Fig. 6).

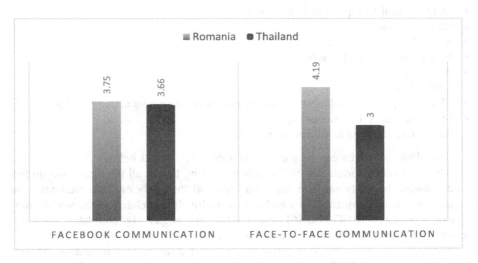

Fig. 6. Facebook versus face-to-face communication

Not only that the difference is significant among the two countries, but, by comparing it to the Facebook conversation frequency and with the results from Fig. 5, it can be concluded that the Thai students are knowing a lot more Facebook friends from the real life, but they are communicating less face-to-face with them, choosing to use more frequently the online social message channel offered by Facebook.

On contrary, for the Romanian students, the things are a little bit the other way around, meaning that they are knowing less Facebook friends from the real life, but they are communicating more face-to-face to their friends. This can also be related to some social aspects, ways of thinking, societal and cultural behaviours etc., which will be analysed in a further research.

Besides this small differences, the other aspects gathered through the questionnaire have shown that both the frequency of connecting and stay connected to Facebook is almost the same: 3.96 for Romanian students to 3.85 for Thai students on a 5-point Likert scale, showing that, on average, a person is connecting a couple of times a day to this social networking platform.

5.4   Comparative Analysis

A number of ten items was constructed based on the questionnaire, eight of these being structured as in Lee's study, while the other two (FNF and FUI) are new and are replacing the PSC (Private Self-Consciousness) and PUSC (Public Self-Consciousness) indicators in the mentioned study. It has been decided to proceed to this modification as these two indicators (namely PSC and PUSC) were not considered so strongly related to the research's purpose and they could easily be integrated in the SCC indicator:

- SCF - Social Comparison Frequency on Facebook;
- SCO - Social Comparison Orientation;
- SE - Self-Esteem;
- SCC - Self-Concept Clarity;
- IU - Intolerance to Uncertainty;
- AXT - Anxiety;
- DPR - Depression;
- FNF - Frequency of a Negative Feeling (when seeing others activity on Facebook);
- FUI - Facebook Use Intensity;
- EXP - Expectations to others' responses.

The obtained results can be seen in the Tables 1, 2 and 3 below.

The determined correlation coefficients are stating that in all the three case studies (Lee's study, the study on Romanian students and the study on Thai students), the social comparison frequency on Facebook is positively correlated to the social comparison orientation (0.470, 0.461 and 0.479) and negatively correlated with the self-esteem level (−0.290, −0.148 and −0.095).

Moreover, a person's self-uncertainty measured through four indicators, namely SCC, IU, AXT and DPR is also positively correlated with the social comparison frequency in both our studies, while in Lee's study the SCC indicator seems to be negatively correlated with the SCF indicator.

Also for the expectation to others' responses, the values obtained through both studies are quite similar, both of them being positively correlated to the social comparison frequency (0.490, 0.521 and 0.541).

As for the PSC and PUSC indicators in Table 1, it can be easily observed that their correlation values with the other indicators are almost the same and they can be gathered, in the future, in a single indicator that can reflect the self-consciousness.

Table 1. Correlation coefficients obtained in Lee's study [12]

	SCF	SCO	SE	SCC	IU	AXT	DPR	PSC	PUSC
SCF	1.000								
SCO	0.470	1.000							
SE	−0.290	−0.070	1.000						
SCC	−0.540	−0.360	0.560	1.000					
IU	0.250	0.180	−0.320	−0.380	1.000				
AXT	0.320	0.160	−0.330	−0.430	0.620	1.000			
DPR	0.310	0.100	−0.560	−0.600	0.380	0.500	1.000		
PSC	0.450	0.460	−0.200	−0.590	0.250	0.250	0.370	1.000	
PUSC	0.450	0.430	−0.280	−0.540	0.300	0.340	0.360	0.600	1.000
EXP	0.490	0.400	−0.080	−0.400	0.290	0.260	0.210	0.330	0.390

Table 2. Correlation coefficients - Romania

	SCF	SCO	SE	SCC	IU	AXT	DPR	FNF	FUI
SCF	1.000								
SCO	0.461	1.000							
SE	−0.148	0.006	1.000						
SCC	0.473	0.394	−0.142	1.000					
IU	0.227	0.345	0.069	0.338	1.000				
AXT	0.247	0.199	−0.036	0.321	0.500	1.000			
DPR	0.353	0.161	−0.162	0.392	0.130	0.239	1.000		
FNF	0.666	0.365	−0.137	0.504	0.243	0.271	0.333	1.000	
FUI	0.414	0.427	0.143	0.378	0.331	0.186	0.140	0.280	1.000
EXP	0.521	0.384	0.093	0.381	0.350	0.262	0.180	0.395	0.522

Table 3. Correlation coefficients - Thailand

	SCF	SCO	SE	SCC	IU	AXT	DPR	FNF	FUI
SCF	1.000								
SCO	0.479	1.000							
SE	−0.095	0.030	1.000						
SCC	0.419	0.264	−0.071	1.000					
IU	0.248	0.351	0.030	0.376	1.000				
AXT	0.295	0.271	0.066	0.393	0.595	1.000			
DPR	0.380	0.156	−0.063	0.424	0.156	0.267	1.000		
FNF	0.690	0.423	−0.098	0.535	0.291	0.285	0.339	1.000	
FUI	0.399	0.418	0.143	0.395	0.334	0.333	0.174	0.320	1.000
EXP	0.541	0.402	0.187	0.291	0.379	0.311	0.183	0.441	0.519

In Tables 2 and 3, these two indicators are mission as they have been incorporated in the SCC indicator, and it can easily be observed that this indicator's value is almost the same with the one obtained in Tables 2 and 3 (0.450 vs. 0.473 in Table 2 and 0.419 in Table 3).

The two new indicators introduced in our study, FNF and FUI, are positively correlated to the social comparison on frequency on Facebook in both studies on Romanian and Thai students, which means that a person that compares frequently with others on Facebook is more probably to feel a negative feeling about herself and also that a person that compares frequently with others on Facebook is more likely to use more often the Facebook page.

As for the SCC concept in comparison with SCO, SE, IU, AXT, DPR the results obtained in the two case studies on Romanian and Thai students are opposite to the one reached in Lee's study: in Table 1 can be detected a negative correlation, while in the Tables 2 and 3 there is a positive one. The cause of this contradiction can be the way in which the questions were addressed in the questionnaire. For example, in our study, one of the questions regarding self-concept clarity was: "In general, my opinions about myself are different from the others" and the answers were 1 for strongly disagree and 5 for strongly agree. The person who answers 1 is, in fact, declaring that his opinion about herself is the same as the others, which means that he has a good self-concept clarity. So, as the grades are decreasing, the self-concept clarity is better, and the correlation between the grades obtained for this question are negative correlated with the ones obtained for the SCO, SE, IU, AXT or DPR.

Having all these observations, it can be said, that, in general, the social comparison is correlated with the other considered variables.

5.5   Grey Analysis

For better shaping the incidence of the social comparison on someone's every-day life, a grey incidence analysis has been proposed between SCO and FNF.

The results obtained for the Romanian students' case are presented in Fig. 7.

Therefore, the synthetic degree of grey incidence among SCO and FNF is 0.61435 which denotes a highly incidence of social comparison on Facebook on the frequency of a negative feeling appearance on each person's state.

Repeating the grey incidence analysis for the Thai students' case, it has been found that the absolute degree of grey incidence is 0.56348, while the relative degree of grey incidence is 0.67226. Therefore, the synthetic degree of grey incidence among SCO and FNF, determined based on both absolute and relative degrees of grey incidence, is 0.61787.

It can be seen that the values obtained for the synthetic degree of grey incidence in both cases are comparable, showing that there is a considerable relationship among the social comparison orientation on Facebook and the frequency of appearance of a negative feeling when watching others activity on this platform.

Even more, by applying an incidence analysis on the number of Facebook friends and the Facebook use intensity, it can be found that the number of friends on Facebook

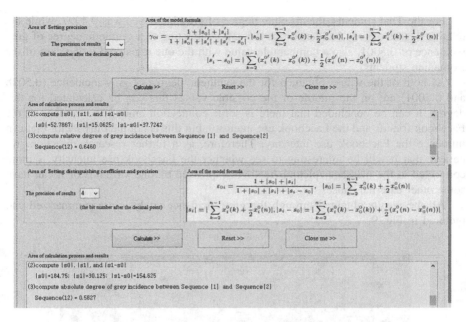

Fig. 7. The relative and absolute degree of grey incidence – SCO vs. FNF - Romania

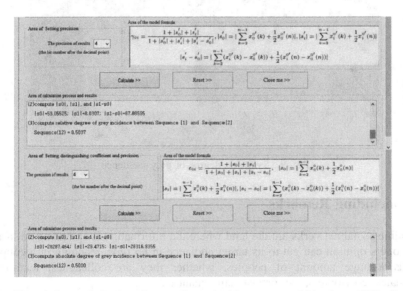

Fig. 8. The relative and absolute degree of grey incidence – no. of friends vs. FUI - Romania

is influencing the usage of this social network by its users as the value determined for the synthetic degree of grey incidence is 0.5018 in the case of Romanian students. The values for the absolute and relative degrees of grey are in Fig. 8.

For the Thai students, the absolute degree of grey incidence was 0.4997, while the relative degree of grey incidence was 0.5004, conducting to a synthetic degree of grey incidence of 0.5001, also a comparable value with the one registered in the case of the Romanian students.

As both of the values obtained for the synthetic degree of grey incidence (0.5018 and 0.5001) are on the middle of the possible values that can be calculated for this degree, it can be concluded that there is some connection among the number of the Facebook friends and the Facebook use intensity, but this is not the only factor that can influence the Facebook use intensity. Therefore, as a further research direction, we believe that it will be interesting to see whether the Facebook use intensity is more connected to the messages/pictures/feelings expressed by one's friends or it can have some other external causes.

All the results obtained through the grey incidence analysis are summarized and pictured in Fig. 9.

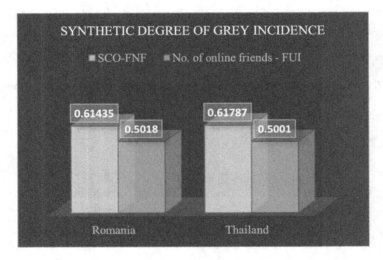

Fig. 9. Synthetic degree of grey incidence

6 Concluding Remarks

The online social networks are the land where the information is spreading so fast and where one's opinion can get to its target audience in no time, reaching in a couple of seconds a huge amount of peoples, whether they are colleagues, team-mates, class-mates, co-workers, etc. or just acquaintances.

But, as this network implies people, the way all this information is transferred from a person to another is very different. Sometimes one can enjoy and be happy for someone else's success, but, in the same time it can be annoyed by another persons' activities, social life, professional life, feelings, traveling experiences, ideas, opinions etc.

Therefore, this paper tries to see whether the people from a randomly chosen sample are comparing themselves with the ones in their own network by considering the posts their friends are making on Facebook (including here the information posted on news feed, their photos, etc.) and whether there is an incidence between the social comparison orientation and the appearance of a negative feeling about themselves. Moreover, the connection between the number of Facebook friends and the frequency of using this social network is analysed. As a result, the correlation analysis is used along with the grey theory incidence analysis.

For conducting this analysis, a study has been made on 143 students from two different countries situated in two different corners of the world and their results have been compared to the ones obtained by Lee in a recent study [12].

The results were conclusive: even though there were some minor differences among the two collected samples, it can be seen that there is a positive correlation between the social comparison orientation and the other analysed factors: social comparison orientation, self-concept clarity, intolerance of uncertainty, anxiety, depression and frequency of a negative feeling, Facebook use intensity and expectations to others' responses.

Even more, the respondents from the both samples of this study have said, that, on average, they are comparing more with others on Facebook when they are visualizing their photos than when they are reading their fiends post. As a future study, it can be tried to see whether the group of friends to whom a person is comparing is compounded by all his Facebook friends or only a certain part of them (colleagues, close friends, co-workers, persons that may have the same background, etc.).

Also, the respondents have declared, on average, that Facebook is an integrating part of their daily routine and that from a certain point they are feeling out of reality when they are not connected.

On average, a person is visiting Facebook a couple times a day. Thus, it has been analysed the incidence on the number of Facebook friends and the Facebook use intensity, and it has been discovered a positively correlation. Also, due to the value determined for the synthetic degree of grey incidence, it has been concluded that not only the number of Facebook friends are influencing a person's Facebook use intensity, but even some other factors related to the quality of their friends posting/advertising/messages/pictures/etc. and, therefore, it needs further investigation in order to determine which one of the other relevant factors may contribute to a higher online activity.

Acknowledgments. This paper was co-financed from the European Social Fund, through the Sectoral Operational Programme Human Resources Development 2007–2013, project number POSDRU/159/1.5/S/138907 "Excellence in scientific interdisciplinary research, doctoral and postdoctoral, in the economic, social and medical fields - EXCELIS", coordinator The Bucharest University of Economic Studies. Also, the authors gratefully acknowledge partial support of this research by Leverhulme Trust International Network research project "IN-2014-020" and by Webster University Thailand.

References

1. Heidemann, J., Klier, M., Probst, F.: Online social networks: a survey of a global phenomenon. Comput. Netw. **56**, 3866–3878 (2012)
2. Statista: http://www.statista.com/statistics/278414/number-of-worldwide-social-network-users/. Accessed December 2014
3. Facebook, Statistics: http://newsroom.fb.com/company-info/. Accessed December 2014
4. Kawamoto, T., Hatano, N.: Viral spreading of daily information in the online social networks. Phys. A **45**, 34–41 (2014)
5. Savage, D., Zhang, X., Yu, X., Chou, P., Wang, Q.: Anomaly detection in online social networks. Soc. Netw. **39**, 62–70 (2014)
6. Zhang, X., You, H., Zhu, W., Qiao, S., Li, J., Gutierrez, L.A., Zhang, Z., Fan, X.: Overlapping community identification approach in online social networks. Phys. A **421**, 233–248 (2015)
7. Xie, J., Kelley, S., Szymanski, B.K.: Overlapping community detection in networks: the state-of-the-art and comparative study. ACM Comput. Surv. **45**(4), 43–51 (2013)
8. Gregory, S.: Fuzzy overlapping communities in networks. J. Stat. Mech: Theory Exp., 20–27 (2011)
9. Goyal, S., Heidari, H., Kearns, M.: Competitive contagion in networks. Games Econ. Behav., 13–22 (2014)
10. Nisan, N., Roughgarden, T., Tardos, E., Vazirani, V.: Algorithmic Game Theory. Cambridge University Press, New York (2007)
11. Sun, Z., Han, L., Huang, W., Wang, X., Zeng, X., Wang, M., Yan, H.: Recommender systems based on social networks. J. Syst. Softw. **99**, 109–119 (2015)
12. Lee, S.Y.: How do people compare themselves with others on social network sites?: The case of Facebook. Comput. Hum. Behav. **32**, 253–260 (2014)
13. Steinfield, C., Ellison, N.B., Lampe, C.: Social capital, self-esteem and use of online social network sites: a longitudinal analysis. J. Appl. Dev. Psychol. **29**(6), 434–445 (2008)
14. Burke, M., Marlow, C. Lento, T.: Social network activity and social well-being. In: Proceeding of the 28th International Conference on Human Factors in Computing Systems, Atlanta, Georgia (2010)
15. Klein, A., Ahf, H., Sharma, V.: Social activity and structural centrality in online social networks. Telematics Inform. **32**, 321–332 (2015)
16. Traud, A., Mucha, P., Porter, M.: Social structure of Facebook networks. Phys. A **391**, 4165–4180 (2012)
17. Girard, Y., Hett, F., Schunk, D.: How individual characteristics shape the structure of social networks. J. Econ. Behav. Organ. **115**, 1–42 (2014)
18. Correa, T., Hinsley, A.W., De Zuniga, H.G.: Who interacts on the web?: The intersection of users' personality and social media use. Comput. Hum. Behav. **26**(2), 247–253 (2010)
19. Bachrach, Y., Kosinski, M., Grapel, T., Kohli, P., Stillwell, D.: Personality and patents of Facebook usage. In: Proceedings of the 3rd Conference ACM Web Science Conference (2012)
20. Mehdizadeh, S.: Self-presentation 2.0.: narcissism and self-esteem on Facebook. Cyberpsychol. Behav. Soc. Netw. **13**(4), 357–364 (2010)
21. Sadovykh, V., Sundaram, D., Piramuthu, S.: Do online social networks support decision-making! Decis. Support Syst. **70**, 15–30 (2015)
22. Bojanowski, M., Corten, R.: Measuring segregation in social networks. Soc. Netw. **39**, 14–32 (2014)

23. Delcea, C.: Not black. Not even white. Definitively grey economic systems. J. Grey Syst. **26**(1), 11–25 (2014)
24. Andrew, A.: Why the world is grey, keynote speech. In: The 3th International Conference IEEE GSIS, Nanjing, China (2011)
25. Forrest, J.: A Systemic Perspective on Cognition and Mathematics. CRC Press, Boca Raton (2013)
26. Sohn, D.: Coping with information in social media: the effects of network structure and knowledge on perception of information value. Comput. Hum. Behav. **32**, 145–151 (2014)
27. Sohn, D.: Disentagling the effects of social network density on electronic word-of-mouth intention. J. Comput. Mediated Commun. **14**(2), 352–367 (2009)
28. Chu, S.C., Choi, S.M.: Electronic word-of-mouth in social networking sites: a cross-cultural study of the United States and China. J. Glob. Mark. **4**(3), 263–281 (2011)
29. Xie, N.-M., Liu, S.-F.: The parallel and uniform properties of several relational models. Syst. Eng. **25**, 98–103 (2007)
30. Liu, S.F., Lin, Y.: Grey Systems Theory and Applications: Understanding Complex Systems. Springer, Heidelberg (2010)
31. Delcea, C., Cotfas, L.-A., Paun, R.: Grey social networks. In: Hwang, D., Jung, J.J., Nguyen, N.-T. (eds.) ICCCI 2014. LNCS, vol. 8733, pp. 125–134. Springer, Heidelberg (2014)

ReproTizer: A Fully Implemented Software Requirements Prioritization Tool

Philip Achimugu, Ali Selamat[✉], and Roliana Ibrahim

Faculty of Computing, Universiti Teknologi Malaysia,
81310 Johor Bahru, Johor, Malaysia
check4philo@gmail.com, {aselamat,roliana}@utm.my

Abstract. Before software is developed, requirements are elicited. These requirements could be over-blown or under-estimated in a way that meeting the expectations of stakeholders becomes a challenge. To develop a software that precisely meets the expectations of stakeholders, elicited requirements need to be prioritized. When requirements are prioritized, contract breaches such as budget over-shoot, exceeding delivery time and missing out important requirements during implementation can be totally avoided. A number of techniques have been developed but these techniques do not addresses some of the crucial issues associated with real-time prioritization of software requirements such as computational complexities and high time consumption rate, inaccurate rank results, inability of dealing with uncertainties or missing weights of requirements, scalability problems and rank update issues. To address these problems, a tool known as ReproTizer (Requirements Prioritizer) is proposed to engender real-time prioritization of software requirements. ReproTizer consist of a WS (Weight Scale) which avails project stakeholders the ability to perceive the influence, different requirements weights may have on the final results. The WS combines a single relative weight decision matrices to determine the weight vectors of requirements with an aggregation operator (AO) which computes the global weights of requirements. The tool was tested for scalability, computational complexity, accuracy, time consumption and rank updates. Results of the performance evaluation showed that the tool is highly reliable (98.89 % accuracy), scalable (prioritized over 1000 requirements), less time consumption and complexity ranging from 500–29,804 milliseconds (ms) of total prioritization time and able to automatically update ranks whenever changes occurs. Requirements prioritization, a multi-criteria decision making task is therefore an integral aspect of the requirements engineering phase of the development life cycle phases. It is used for software release planning and leads to the development of software systems based on the preferential requirements of stakeholders.

Keywords: Software · Requirements · Prioritization · Tool · Stakeholders

Submitted to Transactions on Computational Collective Intelligence (TCCI) Journal.

© Springer-Verlag Berlin Heidelberg 2016
N.T. Nguyen and R. Kowalczyk (Eds.): TCCI XXII, LNCS 9655, pp. 80–105, 2016.
DOI: 10.1007/978-3-662-49619-0_5

1 Introduction

During requirement elicitation, there are more prospective requirements specified for implementation by relevant stakeholders with limited time and resources. Therefore, a meticulously selected set of requirements must be considered for implementation and planning for software releases with respect to available resources. This process is referred to as requirements prioritization. It is considered to be a complex multi-criteria decision making process (Perini et al. 2013).

There are so many advantages of prioritizing requirements before architecture design or coding. Prioritization aids the implementation of a software system with preferential requirements of stakeholders (Ahl 2005; Thakurta 2012). Also, the challenges associated with software development such as limited resources, inadequate budget, insufficient skilled programmers among others makes requirements prioritization really important (Karlsson et al. 2007). It can help in planning for software releases since not all the elicited requirements can be implemented in a single release due to some of these challenges (Berander et al. 2006; Karlsson and Ryan 1997). It also enhances budget control and scheduling (Perini et al. 2013). Therefore, determining which, among a pool of requirements to be implemented first and the order of implementation is necessary to avoid breach of contract or agreement during the development processes. Furthermore, software products that are developed based on prioritized requirements can be expected to have a lower probability of being rejected. To prioritize requirements, stakeholders will have to compare them in order to determine their relative importance through a weight scale which is eventually used to compute the prioritized requirements (Kobayashi and Maekawa 2001). These comparisons becomes complex with increase in the number of requirements (Kassel and Malloy 2003).

Software system's acceptability level is mostly determined by how well the developed system has met or satisfied the specified requirements. Hence, eliciting and prioritizing appropriate requirements and scheduling right releases with the correct functionalities are a critical success factor for building formidable software systems. In other words, when vague or imprecise requirements are implemented, the resulting system will fall short of user's or stakeholder's expectations. Many software development projects have enormous prospective requirements that may be practically impossible to deliver within the expected time frame and budget (Perini et al. 2013; Tonella et al. 2012). It therefore becomes highly necessary to source for appropriate measures for planning and rating requirements in an efficient way.

A number of techniques have been proposed in the literature by authors and scholars, yet many areas of improvement have also been identified to optimize the prioritization processes. With the advent of Internet and quest for software that can service distributed organizations, the number of stakeholders in large-scale projects have drastically increased and requirements are beginning to possess the attributes of evolving due to innovation, technological advancement or business growth. Therefore, prioritization techniques should be able to generate an ordered list of requirements based on the relative weights provided by the project stakeholders at any point during the development life cycle (Perini et al. 2013; Ahl 2005).

The rest of the paper is organized as follows: Sect. 2 discusses the related works while Sect. 3 describes the proposed technique. Section 4 presents an illustrative example of the proposed technique and Sect. 5 describes the attributes of the support tool. Section 6 presents performance evaluation of ReproTizer; Sect. 7 compares the strengths of ReproTizer over existing ones while Sect. 8 concludes the paper and identify areas for future research.

2 Related Work

Many requirements prioritization techniques exist in the literature. All of these techniques utilize a ranking process to prioritize candidate requirements. The ranking process is usually executed by assigning weights across requirements based on pre-defined criteria, such as value of the requirements perceived by relevant stakeholders or the cost of implementing each requirement. From the literature; analytic hierarchy process (AHP) is the most prominently used technique. However, this technique suffers bad scalability. This is due to the fact that, AHP executes ranking by considering the criteria that are defined through an assessment of the relative priorities between pairs of requirements. This becomes impracticable as the number of requirements increases. It also does not support requirements evolution or rank updates but provide efficient or reliable results (Karlsson et al. 1998). Also, all techniques suffer from rank updates issue. This term refers to the inability of a technique to update rank status of ordered requirements whenever a requirement is added or deleted from the list. Prominent techniques that suffer from this limitation are PHandler (Babar et al. 2015), Case base ranking (Perini et al. 2013); Interactive genetic algorithm prioritization technique (Tonella et al. 2012); Binary search tree (Karlsson et al. 1998); Cost value approach (Karlsson and Ryan 1997) and EVOLVE (Greer and Ruhe 2004). Furthermore, existing techniques are prone to computational errors (Ramzan et al. 2011) probably due to lack of robust algorithms. Karlsson et al. (1998) conducted some researches where certain prioritization techniques were empirically evaluated. From their research, they reported that, most of the prioritization techniques apart from AHP and bubble sorts produce unreliable or misleading results while AHP and bubble sorts were also time consuming. The authors then posited that; techniques like hierarchy AHP, spanning tree, binary search tree, priority groups produce unreliable results and are difficult to implement. Babar et al. (2011) were also of the opinion that, techniques like requirement triage, value intelligent prioritization and fuzzy logic based techniques are also error prone due to their reliance on experts and are time consuming too. Planning game has a better variance of numerical computation but suffer from rank updates problem. Wieger's method and requirement triage are relatively acceptable and adoptable by practitioners but these techniques do not support rank updates in the event of requirements evolution as well. Lim and Finkelstein (2012) proposed a method known as StakeRare which stands for Stakeholder Recommender assisted method for requirements elicitation. It is a requirements prioritization method for large projects, where stakeholders can be in different locations and rank requirements based on a 5-point Likert scale. The authors also implemented the concept of StakeRare method into a support tool known as StakeSource2.0 (Lim et al. 2011), which is a web-based tool that supports the StakeRare method. However, the method and tool were not tested for large scale prioritization of requirements. The focus was more on numbers of stakeholders than requirements. Additionally, the proposed approach and tool was not tested with various requirements and scenarios of different organizations.

Our motivation for proposing an improved method and tool arose from the limitations of existing techniques as enumerated below:

(i) Scalability: Techniques like AHP, pairwise comparisons and bubblesort suffer from scalability problems because, requirements are compared based on possible pairs causing n (n − 1)/2 comparisons (Karlsson et al. 1998). For example, when the number of requirements is doubled in a list, other techniques will only require double the effort or time for prioritization while AHP, pairwise comparisons and bubblesort techniques will require four times the effort or time. This is bad scalability.

(ii) Computational complexity: Most of the existing prioritization techniques are actually time consuming in the real world (Karlsson et al. 1998). Ahl (2005) executed a comprehensive experimental evaluation of five different prioritization techniques namely; AHP, binary search tree, planning game, $100 (cumulative voting) and a new method which combines planning game and AHP (PgcAHP), to determine their ease of use, accuracy and scalability. The author went as far as determining the average time taken to prioritize 13 requirements across 14 stakeholders with these techniques. At the end of the experiment; it was observed that, planning game was the fastest while AHP was the slowest. Planning game prioritized 13 requirements in about 2.5 min while AHP prioritized the same number of requirements in about 10.5 min. In other words, planning game technique took only 11.5 s to compute the priority scores of one requirement across 14 stakeholders while AHP consumed 48.5 s to accomplish the same task due to pair comparisons.

(iii) Rank updates: Perini et al. (2013) defined rank update as 'anytime' prioritization; that is, the ability of a technique to automatically update ranks anytime a requirement is included or excluded from the list. This situation has to do with requirements evolution. Therefore, existing prioritization techniques are incapable of updating or reflecting rank status whenever a requirement is introduced or deleted from the rank list. Therefore, it does not support iterative updates. This is very critical because, decision making and selection processes cannot survive without iterations. Therefore, a good and reliable prioritization technique should be one that supports rank updates. This limitation seems to cut across most existing techniques.

(iv) Error proneness: Existing prioritization techniques are also prone to errors (Ramzan et al. 2011). This could be due to the fact that, the rules governing the requirements prioritization processes in the existing techniques are not robust enough. This has also led to the generation of unreliable prioritization results because; such results do not reflect the true ranking of requirements from stakeholder's point of view or assessment after the ranking process. Therefore robust algorithms are required to generate reliable prioritization results.

(v) Lack of fully implemented support tools: From the literature, it was observed that most existing prioritization techniques have not been really implemented for real-life scenarios probably because of the complexities associated with prioritizations and the time required for generating prioritized requirements. Therefore, there is need to implement algorithms that will improve or support requirements prioritization at commercial or industrial level (Peng 2008; Racheva et al. 2008; Ramzan et al. 2009). Before these algorithms can work efficiently, the methods for capturing requirements in an unambiguous way must be well thought of (Grunbacher et al. 2003) since the output of prioritization processes depend on the input and the aim is to plan for software releases (Barney et al. 2006) as well as the successful development of software products in line with negotiated or prioritized requirements (Olson and Rodgers 2002).

3 Proposed Technique

The proposed technique consist of six steps (Fig. 1). The first step is to input the consensus requirements and the criteria describing the expected functionalities of each requirement into ReproTizer. The second step determines the relative value of requirements by indicating the preference weights against requirements using the weight scale (WS) in Table 1. The third step calculates the requirements priority vector, normalize the respective weights and calculate the global weights of requirements (Weight vector). The fourth step elicits the performance of each requirements with respect to the global weights, using a classical weighted average decision matrix (WADM). The fifth and sixth step aggregates and determine the ranks of requirements respectively.

The WS was designed to handle prioritization in both real time and fuzzy conditions. We consider a finite collection of requirements $X = \{R_{11}, R_{12} \ldots R_{1k}\}$ that has to be

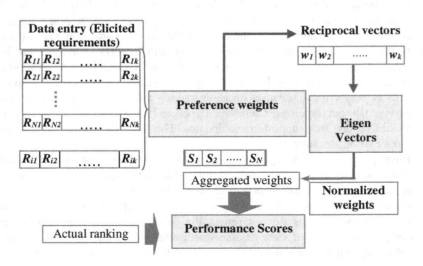

Fig. 1. Proposed technique

ranked against one each other. Our approach consist of set of input R_{11}, R_{12}, ..., R_{1k}, associated with their respective weights w_1, w_2, ..., w_k that represents stakeholders' preferences and a WADM required to calculate the global scores across requirements. The requirement (R_{11} ..., R_{21}..., ..., R_{nk}) represent input data that are ranked using the AO and stored in the database. In this approach, we assume that, the stakeholder's preferences are expressed as relative weights, which are values between 5 and 1.

Table 1. Weight scale (WS)

Terms	Numeric rating	Fuzzy weights
Extremely high (EH)	5	(1,1,1)
Very high (VH)	4	(1/2, 1, 1/3)
High (H)	3	(1/5, 1/2, 1/3)
Fair (F)	2	(1/7, 1/3, 1/5)
Low (L)	1	(1/9, 1/4, 1/7)

The data required for the prioritization process comes from the preference weights of stakeholders which could be imprecise, uncertain and vague due to incomplete information, time limitations, lack of knowledge, or understanding about the system under development. The harmonic mean (HM) is determined to replace requirements with missing weights. This is meant to cater for vagueness associated with require-ments. It is a multi-criteria decision making approach for analyzing the hierarchy of the decision-making process. The proposed approach is used to model the interaction, dependence and feedback within groups of elements and between groups. The groups and elements can be considered as project stakeholders and requirements respectively. Thereafter, the relationships and values between these elements are constructed using a decision matrix. The elements within a group can have a mutual impact on members of the group and the other groups with respect to each of several characteristics. The stakeholder's judgments on the assessment of requirements in the decision-making process always involve incomplete, imprecise, uncertain, intangible and tangible information. Therefore, the conventional approaches seems inadequate to handle the stakeholder's judgments explicitly. To model the uncertainty of stakeholder's relative weights of requirements, harmonic mean computation is integrated into the relative weight scoring process which makes the proposed approach avoid missing weights. The judgment is described through weight numbers where the harmonic mean is used to determine the weights of requirements that were not scored by the stakeholders. Hence, the decision-making process described by the proposed approach is more realistic and capable of generating accurate results.

3.1 Algorithmic Steps of the Computational Process

Step 1: Given a prioritization event E with Requirements R_1, R_2, R_3, ..., R_n (i.e. n – Requirements) and Stakeholders S_1, S_2, S_3, ..., S_u (i.e. u – number of Stakeholders), the

Table 2. Preference weights of requirements

R_1	5	EH	R_{n-1}	R_n
R_2	5	EH		
R_3	5	EH		
R_4	4	VH		

relative or preference weights of requirements are indicated by the project stakeholders as follows:

The weights in Table 2 is for one stakeholders across 4 requirements as an example. For each stakeholder, the proposed approach computes a decision matrix of all the requirements by applying Eq. 1.

$$rank_{j.s_i} \qquad (1)$$

Where $1 \leq j \leq NoOfRequirements$ and
$$1 \leq i \leq NoOfStakeholders$$

Step 2: The sum of the ranks of each requirement is computed across the project stakeholders using Eq. 2.

$$rankSum_j = \sum_{i=1}^{u} rank_j.s_i \qquad (2)$$

Step 3: The reciprocals of the relative weights are determined to minimize the discrepancies of the final ranks by using Eq. 3 and decision matrix is formed as shown in Table 3.

$$reciprocalSum_j = \frac{1}{n} \sum_{i=1}^{u} rank_j.s_i \qquad (3)$$

n stands for the number of requirements undergoing prioritization.

Step 4: The Square of the matrix is computed using Eq. 4 and the sum of each row of the matrix is calculated using Eq. 5 which will yield a result of $(n \times 1)$ matrix, known as the Eigenvector. It represents the global weights of requirements.

$$SquareM = \left(\prod_{i=1}^{n} a_k^2 \right) \qquad (4)$$

$$SumM = \left(\sum_{i=1}^{n} a_k \right) \qquad (5)$$

Table 3. Reciprocals of the preference weights

	S_1	S_2	S_3	...	S_n
R_1	$\frac{1}{n}\sum_{i=1}^{u} rank_1.s_1$	$\frac{1}{n}\sum_{i=1}^{u} rank_1.s_2$	$\frac{1}{n}\sum_{i=1}^{u} rank_1.s_3$	...	$\frac{1}{n}\sum_{i=1}^{u} rank_1.s_{n-1}$
R_2	$\frac{1}{n}\sum_{i=1}^{u} rank_2.s_{2i}$	$\frac{1}{n}\sum_{i=1}^{u} rank_2.s_2$	$\frac{1}{n}\sum_{i=1}^{u} rank_2.s_3$	...	$\frac{1}{n}\sum_{i=1}^{u} rank_1.s_{n-2}$
R_3	$\frac{1}{n}\sum_{i=1}^{u} rank_3.s_3$	$\frac{1}{n}\sum_{i=1}^{u} rank_3.s_2$	$\frac{1}{n}\sum_{i=1}^{u} rank_3.s_3$	...	$\frac{1}{n}\sum_{i=1}^{u} rank_1.s_{n-3}$
...	...		...	...	...
R_n	$\frac{1}{n}\sum_{i=1}^{u} rank_j.s_i$	$\frac{1}{n}\sum_{i=1}^{u} rank_j.s_i$	$\frac{1}{n}\sum_{i=1}^{u} rank_j.s_i$	...	$rankSum_{n-k}$

Step 5: The Eigenvectors are normalized using Eq. 6. Meaning, the sum of all the values in the Eigenvector is calculated and used to divide each of the values in the Eigenvector. This places all the values on a scale of 1 and the sum of all the values to 1.

$$\aleph_j = \frac{w_j}{\sum_{j=1}^{n} w_j = 1} \quad i = 1, \ldots n; j = 1, \ldots m \tag{6}$$

Step 6: This obtains the performance scores for the requirements by summing the relative normalized weights (wj) of each requirement across the stakeholders using Eq. 7.

$$p_i = \sum_{i=1}^{n} w_j \tag{7}$$

4 Illustrative Example

This section presents an illustrative example for prioritizing software requirements with the proposed approach. For the sake of clarity, let us consider 4 requirements to be prioritized by 3 stakeholders. The requirements are *usability*, *scalability*, *security* and *modularity*. Since this is an example, the elicited weights for step 1 is just illustrative and represent opinions of stakeholders. In the implemented tool, the user dialog is achieved with a simplified interface weights scale, shown in Fig. 2 but at the back end, the calculations are performed using the computational processes described in Sect. 3.

It is important to note that in Step 1, stakeholders are only required to provide the preference weights of requirements and the proposed technique automatically perform relevant calculations in order to display the prioritized requirements. Table 4 presents the illustrative preference weights of stakeholders while Table 5 shows the rank sum of weights for the 3 stakeholders using Eq. 2. Table 6 shows the reciprocal values for the rank sum using Eq. 3 and Eq. 4 was used to compute the square matrix of requirements

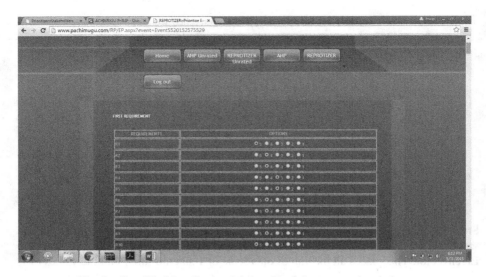

Fig. 2. Simplified interface weights scale of the proposed technique

as displayed in Table 7. The relative normalized decision matrix shown in Table 8 was computed using Eqs. 5 and 6 respectively while final scores for the requirements displayed in Table 9 were computed using Eq. 7. From the final scores of requirements, it can be easily seen that the stakeholders ranked usability and security as the most valued requirements followed by scalability and then modularity. In terms of the accuracy of the proposed approach, it can be seen that the original weights provided by the stakeholders in Table 4 is in agreement with the final scores in Table 9.

Table 4. Preference weights of requirements

	Usability	Scalability	Security	Modularity
Stakeholder 1	5	5	5	5
Stakeholder 2	5	4	5	4
Stakeholder 3	5	4	5	3

Table 5. Rank sum of requirements

	Usability	Scalability	Security	Modularity
Stakeholder 1	15	15	15	15
Stakeholder 2	15	12	15	12
Stakeholder 3	15	12	15	9

Table 6. Reciprocal values of the requirements' sum

	Usability	Scalability	Security	Modularity
Stakeholder 1	3.75	3.75	3.75	3.75
Stakeholder 2	3.75	3.00	3.75	3.00
Stakeholder 3	3.75	3.00	3.75	2.25

Table 7. Square matrix of the requirements' sum

	Usability	Scalability	Security	Modularity
Stakeholder 1	14.06	14.06	14.06	14.06
Stakeholder 2	14.06	9.00	14.06	9.00
Stakeholder 3	14.06	9.00	14.06	5.06

Table 8. Normalized weights

	Usability	Scalability	Security	Modularity
Stakeholder 1	0.304	0.304	0.304	0.304
Stakeholder 2	0.305	0.195	0.305	0.195
Stakeholder 3	0.333	0.213	0.333	0.120

Table 9. Performance scores

Requirements	Final Scores
Usability	0.942
Scalability	0.712
Security	0.942
Modularity	0.619

5 Tool Support

The tool was implemented in C#, very similar to Java platform standard edition 7. It takes relative weights of requirements provided by the stakeholders as input and processes them to generate list of prioritized requirements. The tool is deployed at http://www.pachimugu.com/. It provides a convenient way of accessing various menus of the tool from the HTML of the page. Additionally, a pattern matching was utilized to aid the re-weighting and re-computation of ranks whenever requirements evolves. The tool also provides an avenue for inclusion or exclusion of stakeholders if need be using three step process; (1) New stakeholders are added by the administrator as soon as they get registered as users. The proposed tool can cater for as much stakeholders as required for a particular software project. (2) The consensus requirements automatically appears against their names so as to initiate the scoring process. (3) The relative weights are then processed or computed to display the final ranks of requirements. However, deleting a stakeholder also applies to the relative weights of that stakeholder where the tool automatically re-compute the new ranks of each requirement based on the new number of stakeholders. The tool's main window is displayed in Fig. 3.

Considering the top-most part of the window, it can be observed that the name of the tool is known as Requirements Prioritizer, consisting of five tabs namely; *Home*, *Events*, *Login*, *Sign Up* and *Contact*. To use this tool, prospective project stakeholders would have to first register by clicking the *sign up* tab to fill the required details. Once this is done, the tools' administrator can now view all the registered stakeholders and

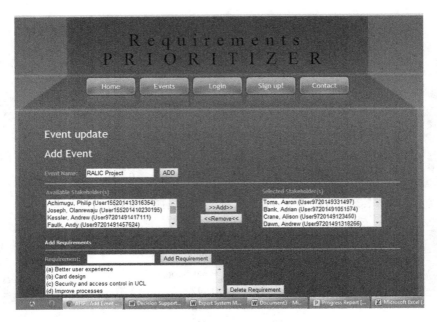

Fig. 3. Proposed tool main window

add them up. It is also the duty of the administrator to input the elicited requirements to undergo prioritization into ReproTizer. Requirements are inputted into ReproTizer by clicking the tab *add event* where the name of the project is used to save the inputted requirements. It can be observed from the window that the tool is flexible enough to cater for addition or deletion of requirements or stakeholders at any point in time where ReproTizer simply updates the ranks status of requirements by displaying new ordered list of requirements that has occurred either by adding or deleting a requirement or stakeholder. A concept Perini et al. described as "anytime prioritization" (Perini et al. 2013). Once, all the requirements have been inputted into ReproTizer and all the registered project stakeholders have been accepted by the administrator, the scoring of requirements can be initiated by stakeholders who logs into ReproTizer with their respective username and password. Figure 4 shows a window of the database where the registered stakeholders are stored.

Once the stakeholders log into ReproTizer, they can now view the consensus requirements in order to rank or score them. Figure 5 presents the window that shows the individual weights of stakeholders. The assessment of these requirements lead to the construction of a decision matrix. This is where the tradeoffs between the requirements are displayed. ReproTizer displays both the individual and overall ratings of each requirements. The individual weights signifies the ranks of the requirements by one stakeholder. ReproTizer automatically calculates the overall weights of requirements by aggregating the scores across all project stakeholders in chronological order (Fig. 6). If the requirements weights are inconsistent or missing, a message pops-up warning the user.

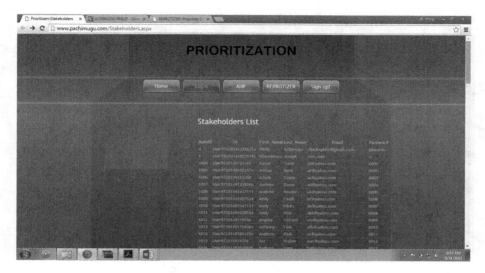

Fig. 4. Registered stakeholders

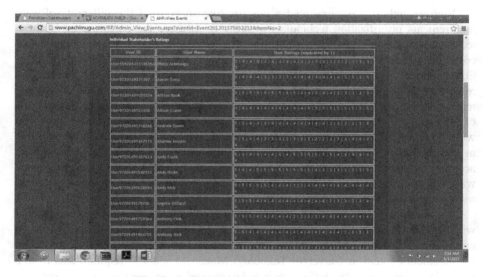

Fig. 5. Individual weights of requirements

6 Performance Evaluation

The motivation for developing ReproTizer was as a result of the following limitations of existing techniques as described in Sect. 2 (scalability, computational complexity, rank updates, error proneness and lack of fully implemented support tools). Therefore, the evaluation of ReproTizer is based on these parameters.

Fig. 6. Overall weights of requirements (Final ranks)

Various authors have executed a comparative analysis of the different software requirements prioritization techniques in order to measure the performance of these techniques. In this section, some well-known requirements prioritization techniques are considered and compared with ReproTizer based on the five evaluation criteria mentioned above. Consequently, *scalability* is measured in terms of the number of requirements ReproTizer can accommodate at runtime. *Computational complexity* measures the time consumed in executing the computational processes or calculations of the weighted requirements to generate the prioritized list. *Rank updates* has to do with the ability of ReproTizer to effect or generate new ranks whenever a requirement or stakeholder is included or excluded from the list. *Error proneness* measures the accuracy of the ranked results while lack of fully implemented support tools has to do with the absence of tool capable of supporting real-time prioritization of software requirements.

In order to evaluate the performance of ReproTizer, 4 experiments were conducted with different requirements datasets. The first experiment was conducted with 20 requirements from GSMS project (A web-based Graduate Students' information Management System in Universiti Teknologi Malaysia) and 100 requirements from a health information system (HIS) software. The second experiment was conducted with 200 requirements from RALIC project (an access/identity card software for university staff and students in University College London). The third and fourth experiment were conducted with 500 and then, 1000 requirements of an enterprise resource planning (ERP) software package. These experiments were meant to prove the contributions of ReproTizer with respect to the limitations highlighted in Table 10. As it can be seen, a lot of techniques suffer scalability problems. Most techniques are only suitable for small to medium sized software projects. To address scalability issues, Babar and colleagues proposed an expert system known as PHandler which was able to prioritize up to 500 requirements; the highest so far in the literature (Babar et al. 2015). However, if requirements run up to thousands, it is not certain that PHandler can provide desired results on that scale. PHandler was not also tested for computational complexities, rank updates and time consumption; although, their system was only meant to address scalability issue inherent in existing techniques. In terms of time consumption, a

Table 10. Limitations of existing techniques

Techniques and references	Limitations
AHP (Saaty 1980; Karlsson et al. 1998), Binary tree (Beg et al. 2009; Aasem et al. 2010), Case based ranking (Perini et al. 2013), Interactive requirements prioritization (Tonella et al. 2013), Cost-Value Ranking (Karlsson and Ryan 1997), StakeSource2.0 (Lim et al. 2011), Fuzzy AHP (Lima et al. 2011), Quality Functional Deployment (QFD) (Edwin 1992), Ranking, Requirement uncertainty prioritization approach (RUPA) (Voola and Babu 2012), Round-the-Group Prioritization (Hatton 2008; Karlsson and Ryan 1997), $100 Allocation or Cumulative Voting (Berander and Andrews 2005; Regnell et al. 2001)	Not scalable, 10–100 requirements only
Cost-Value Ranking (Karlsson and Ryan 1997), AHP (Saaty 1980; Karlsson et al. 1998), Binary search tree (Duan et al. 2009)	Time consuming
EVOLVE (Thakurta 2013, Greer and Ruhe 2004), Wiegers' matrix approach (Duan et al. 2009)	Computationally complex
Hierarchy AHP (Karlsson et al. 1998), Minimal spanning tree (Karlsson et al. 1998), Multi-criteria Preference Analysis Requirements Negotiation (MPARN) (In and Olson 2002), Pair Wise Analysis (Karlsson and Ryan 1997), Quality Functional Deployment (QFD) (QFD) (Edwin 1992), Simple multi-criteria rating technique by swing (SMARTS) (Avesani et al. 2005), Top ten requirements (Berander 2004), Value based requirements prioritization (Kukreja et al. 2012), WinWin (Gruenbacher 2000)	Error prone
TOPSIS (Kukreja 2013; Kukreja et al. 2012), Requirements triage (Karlsson et al. 2004), PHandler (Babar et al. 2015)	Lack of implemented tool, do not recall or update ranks and time consumption rate was not measured

number of techniques are also limited in this area. Some studies confirmed that most techniques are time consuming (Ramzan et al. 2011; Soni 2014; Kyosev 2014; Dabbagh and Lee 2014). Specifically, AHP, cumulative voting, numerical assignment, ranking, top-ten, Theory-W, planning game, requirements triage, Wieger's method and value based requirement prioritization techniques consumes a lot of time during the prioritization process (Ramzan et al. 2011). Furthermore, the systematic literature review executed by Achimugu et al. (2014) have it that most techniques suffer from rank inaccuracies, computational complexities, rank updates, scalability, requirements dependencies among others.

Requirements for software projects 1–4 were inputted into ReproTizer. This was followed by the indication of preference weights against each requirements where ReproTizer was automatically able to display prioritized requirements based on the individual and overall weights of requirements. We have observed that results of prioritization often get faulty when requirements increases due to computational complexities and lack of efficient algorithms. However, in the case of ReproTizer, Figs. 7a and 7b show the average accuracy and time consumed for prioritizing 20 requirements while Fig. 7c shows that ReproTizer is automatically able to update rank status when requirements evolves. Similarly, Figs. 8a, 8b; 9a, 9b and 10a, 10b show the average accuracy and time consumed by ReproTizer for prioritizing 200, 500 and 1000 requirements respectively while Figs. 8c, 9c and 10c confirmed that ReproTizer is capable of updating rank status when requirements changes on a large scale. For the time consumption, it took ReproTizer 0.39 min (23.4 s) to prioritize 500 requirements (Fig. 9b) while 0.49 min (29.4 s) was exhausted in prioritizing 1000 requirement (Fig. 10b). The time difference between prioritizing 500 and 1000 requirements is 1 min which is expected because the requirements are doubled. This would almost mean that, for every 500 requirements; 1 additional minute is consumed by ReproTizer to produce the desired results. This is good response time achieved by implementing improved formulas and algorithms Therefore, we conclude that, a fully implemented support tool with high accuracy, good response time and user-friendlier interface for software requirements prioritization has been developed. Also, using a six-step approach, ReproTizer is able to automatically calculate the weights of requirements and perform trade-offs in all steps with minimized divergence in prioritized requirements.

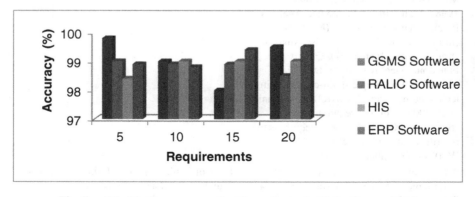

Fig. 7a. Prioritization accuracy for 20 requirements (Color figure online)

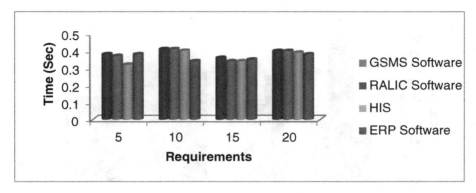

Fig. 7b. Time taken for prioritizing 20 requirements (Color figure online)

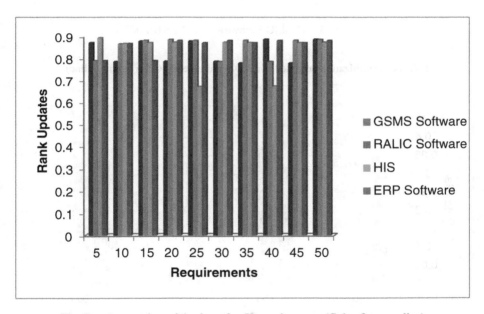

Fig. 7c. Automatic rank updates for 50 requirements (Color figure online)

All our experiments were carried out on a computer with a 2.4 GHz processor and 4 GB RAM. We have observed that ReproTizer consumed an average run time ranging from 500–29,804 milliseconds (ms) to prioritize requirements on a large scale. Table 11 shows the average runtime of three major components that constitute ReproTizer. The average runtime for the decision matrix includes time taken for the construction of preference weights of requirements. Similarly, average runtime for computing the normalized decision matrix includes time taken for constructing a new matrix which subjects the summation of all the preference weights of a requirement to 1 while the global decision matrix stands for the average time of computing the final

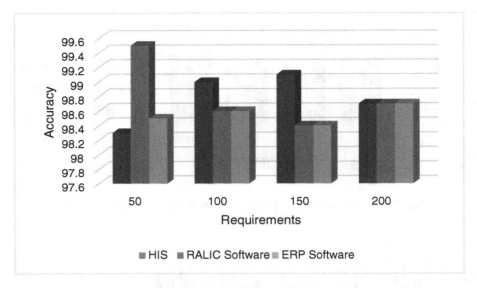

Fig. 8a. Prioritization accuracy for 200 requirements (Color figure online)

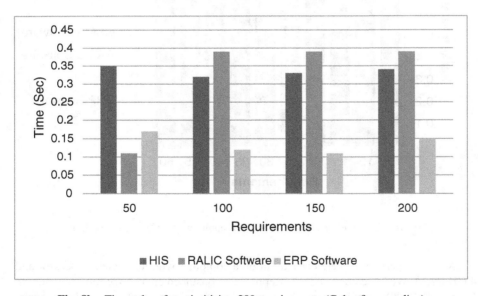

Fig. 8b. Time taken for prioritizing 200 requirements (Color figure online)

weights of requirements. Among these three modules, the discrepancy rate is highly minimal with high correlation between the relative and final weights. Therefore, ReproTizer produces good response time with reduced complexities.

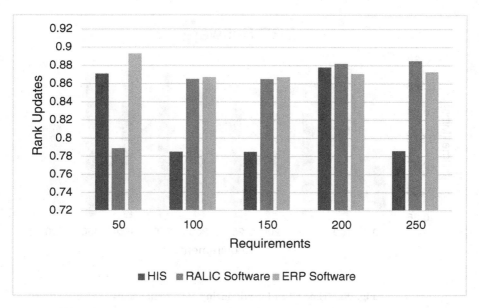

Fig. 8c. Automatic rank updates for 250 requirements (Color figure online)

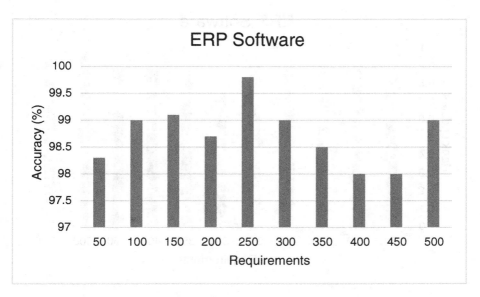

Fig. 9a. Prioritization accuracy for 500 requirements

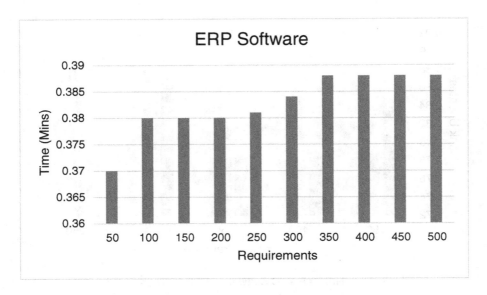

Fig. 9b. Time taken for prioritizing 500 requirements

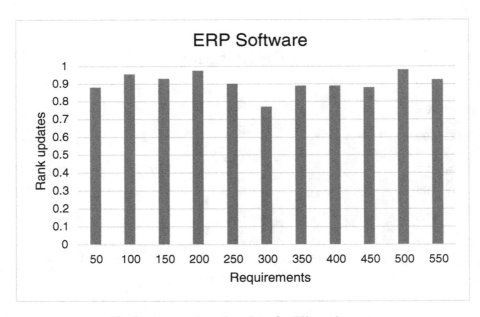

Fig. 9c. Automatic rank updates for 550 requirements

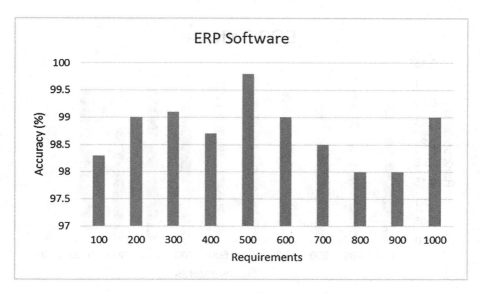

Fig. 10a. Prioritization accuracy for 1000 requirements

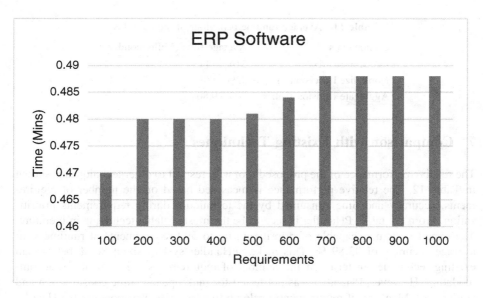

Fig. 10b. Time taken for prioritizing 1000 requirements

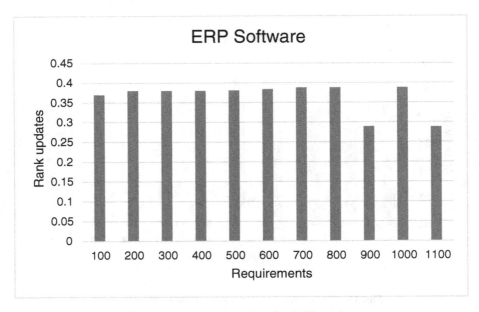

Fig. 10c. Automatic rank updates for 1100 requirements

Table 11. Average runtime behaviour of the modules

Components	Average time (Milliseconds)
Decision matrix	3192–3443
Normalized decision matrix	2290–894
Aggregate decision matrix	500–29,804

7 Comparison with Existing Techniques

The relative performance of the proposed tool with respect to other techniques is shown in Table 12. The relative performance is measured based on the number of requirements, accuracy and time consumed by the techniques during requirements prioritization. From the table, PHandler is seen to be the most scalable technique in literature. The expert system is capable of prioritizing up to 500 requirements at runtime with average accuracy of 93.89 %. This makes PHandler system about 80 % better than existing techniques in terms of the number of requirements it is capable of accommodating. However, PHandler was not tested for rank updates, time consumption and complexity. Meaning, if requirements scale up to thousands, it is not sure if PHandler would produce the desired results at that scale. This forms the rationale for developing a support tool capable of prioritizing more numbers of requirements. Hence, Repro-Tizer was developed and evaluated with 5 different software projects requirements ranging from small, medium and to large scale. The average accuracy of ReproTizer was 98.89 % even on a large scale. It was also able to accommodate and prioritize over

1000 requirements with less complexity between 500–29,804 ms thereby producing good response time. When compared to other techniques in literature, the capabilities of ReproTizer are eminent. Besides, the computational processes, formulas and algorithm are simple but robust enough to be used in practice. The tool has been fully implemented and deployed online, available for use by software practitioners on real-life basis. The performance of the proposed tool was generally evaluated based on number of requirements, time consumption and computational complexities and rank updates. Based on these evaluation parameters, it is clear that ReproTizer is much better and would be beneficial in practice.

Table 12. Comparative analysis of prioritization techniques.

Source	Technique	No of requirements	Accuracy	Time consumption	Support tool
(Ramzan et al. 2011)	Intelligent requirement prioritization	Not indicated	90 %	90 Work hours	×
(Ramzan et al. 2011)	Theory W	Not indicated	80 %	160 Work hours	×
(Perini et al. 2009)	AHP	20	85 %	37 min	√
(Ramzan et al. 2011)	Cumulative voting	Not indicated	85 %	120 Work hours	×
(Ramzan et al. 2011)	Wieger's method	Not indicated	85 %	100 Work hours	×
(Perini et al. 2009)	Case-Based Ranking	20	Not indicated	10 min	√
(Perini et al. 2013)	Case-Based Ranking	25, 50, 100	80 %	Not measured	×
(Tonella et al. 2012)	Interactive GA-Based Prioritization	26, 23, 21 and 49	97.20 %	Not measured	×
(Lim and Finkelstein 2012; Lim et al. 2011)	StakeRare, StakeSource 2.0	<50	80 %	Not measured	√
(Babar et al. 2015)	PHandler	14, 25, 50, 100, 200, 400, 500	93.89	Not measured	×
This Study	ReproTizer	20, 50, 100, 200, 500, 1000	98.89 %	500– 29,804 ms (0.5– 29.804 s)	√

8   Conclusion/Future Work

The aim of this research was to identify the limitations of existing prioritization techniques so as to address them. It was eventually discovered that existing techniques actually suffer from mainly scalability problems, large disparity or disagreement between ranked weights, rank reversals, as well as unreliable results. These were all taken into cognizance during the course of developing ReproTizer. The method utilized in this research consisted of intelligent algorithms implemented with C# and Micro-softSQL server 2012. Efficient models were formulated in order to enhance the reliability of the proposed approach. The developed tool was designed and implemented to cater large requirements and stakeholders. It is easy to use with friendlier user interface, reduced computational complexities and has addressed rank reversals issues. For the future work, we hope to validate the tool in a real-life setting with large numbers of stakeholders and requirements alike. Finally, the developed tool is able to classify ranked requirements in chronological order with an accompanied graph to visualize the prioritized results at a glance. For dependency issues, requirements are thoroughly analyzed using factor analysis to track redundant, conflicting, independent and dependent requirements before inputting the requirements into ReproTizer.

Acknowledgement. The Universiti Teknologi Malaysia (UTM) under Research University funding vot number 02G31 and Ministry of Higher Education (MOHE) Malaysia under vot number 4F550 are hereby sincerely acknowledged for providing the research funds to complete this research.

References

Perini, A., Ricca, F., Susi, A., Bazzanella, C.: An empirical study to compare the accuracy of AHP and CBRanking techniques for requirements prioritization. In: Proceedings of the Fifth International Workshop on Comparative Evaluation in Requirements Engineering, pp. 23–35. IEEE (2007)

Ruhe, G., Eberlein, A., Pfahl, D.: Trade-off analysis for requirements selection. Int. J. Softw. Eng. Knowl. Eng. **13**(4), 345–366 (2003)

Ahl, V.: An experimental comparison of five prioritization methods–investigating ease of use, accuracy and scalability. Master's thesis, School of Engineering, Blekinge Institute of Technology, Sweden, August 2005

Berander, P., Khan, K.A., Lehtola, L.: Towards a research framework on requirements prioritization. In: Proceedings of Sixth Conference on Software Engineering Research and Practice in Sweden (SERPS 2006), October 2006

Kobayashi, M., Maekawa, M.: Need-based requirements change management. In: Proceedings of ECBS 2001 Eighth Annual IEEE International Conference and Workshop on the Engineering of Computer Based Systems, pp. 171–178 (2001)

Kassel, N.W., Malloy, B.A.: An approach to automate requirements elicitation and specification. In: Proceedings of the 7th IASTED International Conference on Software Engineering and Applications, Marina Del Rey, CA, USA, 3–5 November 2003

Perini, A., Susi, A., Avesani, P.: A machine learning approach to software requirements prioritization. IEEE Trans. Softw. Eng. **39**(4), 445–460 (2013)

Tonella, P., Susi, A., Palma, F.: Interactive requirements prioritization using a genetic algorithm. Inf. Softw. Technol. Inf. Softw. Technol. **55**, 173–187 (2012)

Babar, M.I., Ghazali, M., Jawawi, D.N., Shamsuddin, S.M., Ibrahim, N.: PHandler: an expert system for a scalable software requirements prioritization process. Knowl.-Based Syst. **84**, 179–202 (2015)

Kaur, G., Bawa, S.: A survey of requirement prioritization methods. Int. J. Eng. Res. Technol. **2**(5), 958–962 (2013)

Voola, P., Babu, A.: Requirements uncertainty prioritization approach: a novel approach for requirements prioritization. Softw. Eng. Int. J. (SEIJ) **2**(2), 37–49 (2012)

Thakurta, R.: A framework for prioritization of quality requirements for inclusion in a software project. Softw. Qual. J. **21**, 573–597 (2012)

Ramzan, M., Jaffar, A., Shahid, A.: Value based intelligent requirement prioritization (VIRP): expert driven fuzzy logic based prioritization technique. Int. J. Innovative Comput. **7**(3), 1017–1038 (2011)

Perini, A., Ricca, F., Susi, A.: Tool-supported requirements prioritization: comparing the AHP and CBRank method. Inf. Softw. Technol. **51**, 1021–1032 (2009)

Greer, D., Ruhe, G.: Software release planning: an evolutionary and iterative approach. Inf. Softw. Technol. **46**(4), 243–253 (2004)

Franceschini, F., Rupil, A.: Rating scales and prioritization in QFD. Int. J. Qual. Reliab. Manage. **16**(1), 85–97 (1999)

Karlsson, J., Wohlin, C., Regnell, B.: An evaluation of methods for prioritizing software requirements. Inf. Softw. Technol. **39**(14), 939–947 (1998)

Kukreja, N., Payyavula, S., Boehm, B., Padmanabhuni, S.: Value-based requirements prioritization: usage experiences. Procedia Comput. Sci. **16**, 806–813 (2012)

Kukreja, N.: Decision theoretic requirements prioritization: a two-step approach for sliding towards value realization. In: Proceedings of the 2013 International Conference on Software Engineering, pp. 1465–1467. IEEE Press (2013)

Dabbagh, M., Lee, S.: An approach for integrating the prioritization of functional and nonfunctional requirements. Sci. World J. (2014)

Voola, P., Vinaya Babu, A.: Interval evidential reasoning algorithm for requirements prioritization. In: Satapathy, S.C., Avadhani, P.S., Abraham, A. (eds.) Proceedings of the InConINDIA 2012. AISC, vol. 132, pp. 915–922. Springer, Heidelberg (2012)

Aasem, M., Ramzan, M., Jaffar, A.: Analysis and optimization of software requirements prioritization techniques. In: 2010 International Conference on Information and Emerging Technologies (ICIET), pp. 1–6. IEEE (2010)

Racheva, Z., Daneva, M., Herrmann, A., Wieringa, R.: A conceptual model and process for client-driven agile requirements prioritization. In: 2010 Fourth International Conference on Research Challenges in Information Science (RCIS), pp. 287–298. IEEE (2010)

Otero, C., Dell, E., Qureshi, A., Otero, L.: A quality-based requirement prioritization framework using binary inputs. In: 2010 Fourth Asia International Conference on Mathematical/Analytical Modelling and Computer Simulation (AMS), pp. 187–192. IEEE (2010)

Carod, N., Cechich, A.: Cognitive-driven requirements prioritization: a case study. In: 2010 9th IEEE International Conference on Cognitive Informatics (ICCI), pp. 75–82. IEEE (2010)

Gaur, V., Soni, A., Bedi, P.: An agent-oriented approach to requirements engineering. In: 2010 IEEE 2nd International Advance Computing Conference (IACC), pp. 449–454 (2010)

Beg, M., Verma, R., Joshi, A.: Reduction in number of comparisons for requirement prioritization using B-Tree. In: IEEE International Advance Computing Conference, 2009, IACC 2009, pp. 340–344. IEEE (2009)

Hatton, S.: Choosing the right prioritisation method. In: 19th Australian Conference on Software Engineering, 2008, ASWEC 2008, pp. 517–526. IEEE (2008)

Daneva, M., Herrmann, A.: Requirements prioritization based on benefit and cost prediction: a method classification framework. In: EUROMICRO-SEAA, pp. 240–247. IEEE (2008a)

Beg, R., Abbas, Q., Verma, R.P.: An approach for requirement prioritization using b-tree. In: First International Conference on Emerging Trends in Engineering and Technology, 2008, ICETET 2008, pp. 1216–1221. IEEE (2008)

Laurent, P., Cleland-Huang, J., Duan, C.: Towards automated requirements triage. In: 15th IEEE International Requirements Engineering Conference, 2007, RE 2007, pp. 131–140 (2007)

Avesani, P., Bazzanella, C., Perini, A., Susi, A.: Facing scalability issues in requirements prioritization with machine learning techniques. In: RE 2005, pp. 297–306 (2005)

Avesani, P., Bazzanella, C., Perini, A., Susi, A.: Supporting the requirements prioritization process: a machine learning approach. In: Proceedings of 16th International Conference on Software Engineering and Knowledge Engineering, SEKE 2004, pp. 306–311. KSI Press, Banff (2004)

Moisiadis, F.: The Fundamentals of prioritizing requirements. In: Proceedings of Systems Engineering Test and Evaluation Conference, SETE 2002 (2002)

Aaron, K.M., Paul, N., Anton, A.I.: Prioritizing legal requirements. In: Second International Workshop on Requirements Engineering and Law, 2009, RELAW 2009, pp. 27–32. IEEE (2009)

Svahnberg, M., Karasira, A.: A study on the importance of order in requirements prioritisation. In: 2009 Third International Workshop on Software Product Management (IWSPM), pp. 35–41. IEEE (2009)

Tonella, P., Susi, A., Palma, F.: Using interactive GA for requirements prioritization. In: 2010 Second International Symposium on Search Based Software Engineering (SSBSE), pp. 57–66. IEEE (2010)

Bebensee, T., van de Weerd, I., Brinkkemper, S.: Binary priority list for prioritizing software requirements. In: Wieringa, R., Persson, A. (eds.) REFSQ 2010. LNCS, vol. 6182, pp. 67–78. Springer, Heidelberg (2010)

Duan, C., Laurent, P., Cleland-Huang, J., Kwiatkowski, C.: Towards automated requirements prioritization and triage. Requir. Eng. 14(2), 73–89 (2009)

Carod, N., Cechich, A.: Requirements Prioritization Techniques (2001)

Karlsson, L., Thelin, T., Regnell, B., Berander, P., Wohlin, C.: Pair-wise comparisons versus planning game partitioning-experiments on requirements prioritisation techniques. Empir. Softw. Eng. 12(1), 3–33 (2007)

Lehtola, L., Kauppinen, M.: Suitability of requirements prioritization methods for market-driven software product development. Softw. Process Improv. Pract. 11(1), 7–19 (2006)

Berander, P., Andrews, A.: Requirements prioritization. In: Aurum, A., Wohlin, C. (eds.) Engineering and Managing Software Requirements, pp. 69–94. Springer, Heidelberg (2005)

Lehtola, L., Kauppinen, M., Kujala, S.: Requirements prioritization challenges in practice. In: Bomarius, F., Iida, H. (eds.) PROFES 2004. LNCS, vol. 3009, pp. 497–508. Springer, Heidelberg (2004)

Karlsson, J., Ryan, K.: A cost-value approach for prioritizing requirements. IEEE Softw. 14, 67–74 (1997)

Saaty, T.L.: The Analytic Hierarchy Process. McGraw-Hill, New York (1980)

Herrmann, A., Daneva, M.: Requirements prioritization based on benefit and cost prediction: an agenda for future research. In: RE 2008, pp. 125–134. IEEE Computer Society (2008b)

Wiegers, K.E.: First things first: prioritizing requirements. Softw. Dev. 7(9) (1999). www.processimpact.com/pubs.shtml#requirements

Regnell, B., Host, M., Dag, J.: An industrial case study on distributed prioritization in market-driven requirements engineering for packaged software. Requir. Eng. 6, 51–62 (2001)

Edwin, D.: Quality function deployment for large systems. In: International Engineering Management Conference 1992, Eatontown, NJ, USA, 25–28 October 1992

Olson, H., Rodgers, T.: Multi-criteria preference analysis for systematic requirements negotiation. In: COMPSAC 2002, pp. 887–892 (2002)

Berander, P.: Prioritization of Stakeholder Needs in Software Engineering. Understanding and Evaluation. Licenciate Thesis, Blekinge Institute of Technology, Sweden, Licentiate Series, 12 (2004)

Karlsson, J., Olsson, S., Ryan, K.: Improved practical support for large scale requirements prioritizing. J. Requir. Eng. **2**, 51–67 (1997)

Peng, S.: Sample selection: an algorithm for requirements prioritization. ACM (2008)

Racheva, Z., Daneva, M., Buglione, L.: Supporting the dynamic reprioritization of requirements in agile development of software products. In: Second International Workshop on Software Product Management, 2008, IWSPM 2008, pp. 49–58. IEEE (2008)

Lim, S.L., Finkelstein, A.: StakeRare: using social networks and collaborative filtering for large-scale requirements elicitation. IEEE Trans. Softw. Eng. **38**(3), 707–735 (2012)

Kyosev, T.H.: Comparing Requirements Prioritization Methods in Industry: A study of the Effectiveness of the Ranking Method, the Binary Search Tree Method and the Wiegers Matrix. MSc Thesis, Negometrix BV, Germany (2014)

Babar, M., Ramzan, M., Ghayyur, S.: Challenges and future trends in software requirements prioritization. In: 2011 International Conference on Computer Networks and Information Technology (ICCNIT), pp. 319–324. IEEE (2011)

Gruenbacher, P.: Collaborative requirements negotiation with easy winwin. In: Proceedings of 2nd International Workshop on the Requirements Engineering Process, Greenwich London, September 2000

Lima, D.C., Freitas, F., Campos, G., Souza, J.: A fuzzy approach to requirements prioritization. In: Cohen, M.B., Ó Cinnéide, M. (eds.) SSBSE 2011. LNCS, vol. 6956, pp. 64–69. Springer, Heidelberg (2011)

Barney, S., Aurum, A., Wohlin, C.: Quest for a silver bullet: creating software product value through requirements selection. In: 32nd EUROMICRO Conference on Software Engineering and Advanced Applications, SEAA 2006. pp. 274–281. IEEE (2006)

Karlsson, L., Berander, P., Regnell, B., Wohlin,C.: Requirements prioritization: an experiment on exhaustive pair wise comparisons versus planning game partitioning. In: Proceedings of Empirical Assessment in Software Engineering (EASE 2004), Edinburgh, Scotland (2004)

Grunbacher, P., Halling, M., Biffl, S., Kitapci, H., Boehm, B.: Repeatable quality assurance techniques for requirements negotiations. In: Proceedings of the 36th Annual Hawaii International Conference on System Sciences, 9 p. IEEE (2003)

Ramzan, M., Arfan, J., AlIliad, I., Anwar, S., Shahid, A.: Value based fuzzy requirement prioritization and its evaluation framework. In: Fourth International Conference on Innovative Computing, Information and Control. pp. 1464–1468 (2009)

Achimugu, P., Selamat, A., Ibrahim, R., Mahrin, M.N.R.: A systematic literature review of software requirements prioritization research. Inf. Softw. Technol. **56**(6), 568–585 (2014)

Lim, S.L., Damian, D., Finkelstein, A.: StakeSource 2.0: using social networks of stakeholders to identify and prioritise requirements. In: Proceedings of the 33rd International Conference on Software Engineering, pp. 1022–1024. ACM (2011)

Soni, A.: An evaluation of requirements prioritisation methods. Int. J. Innovative Res. Adv. Eng. **1**(10), 402–411 (2014)

A Consensus-Based Method for Solving Concept-Level Conflict in Ontology Integration

Trung Van Nguyen[1] and Hanh Huu Hoang[2]([✉])

[1] College of Sciences, Hue University, 77 Nguyen Hue Street, Hue City, Vietnam
nvtrung@hueuni.edu.vn
[2] Hue University, 3 Le Loi Street, Hue City, Vietnam
hhhanh@hueuni.edu.vn

Abstract. Ontology reuse has played an important role in developing the shared knowledge in Semantic Web. The ontology reuse enables knowledge sharing more easily between ontology-based intelligent systems. In meanwhile, we are still facing the challenging task of solving the conflict potentials in the ontology integration at syntactic and semantic levels. On one aspect of considering knowledge conflicts during the integration process, we try to find the meaningfulness of the conflicting knowledge that means a consensus among conflicts in integrating ontologies. This paper presents a novelty method for finding the consensus in ontology integration at the concept level. Our approach is based on the consensus theory and distance functions between attributes' values.

Keywords: Consensus theory · Ontology integration · Concept level · Similarity distance

1 Introduction

Ontology integration is a vital problem in ontological management and engineering for knowledge sharing and reuse. The ontology reuse has been a key factor in ontology development in terms of enabling knowledge sharing between ontology-based intelligent systems. For instance, we would like to build an ontology for a semantic- and knowledge-based system, we can start of referencing common ontologies or even using existing knowledge in datasets from the Linked Open Data (LOD) cloud. Let's take a look at mOntage framework [3] which supports domain experts defining a domain ontology schema and automatically populating the ontology with instances obtained from selected sources of LOD cloud. The most important challenge for solutions like mOntage is the conceptual heterogeneity, which is also called semantic heterogeneity [5] and logical mismatch [10]. These problems occur due to the use of different axioms for defining concepts or due to the use of totally different concepts. For instance, in datasets, we can have airport instances which are described with the type of dbpedia-owl:Airport that is a subClass of dbpedia-owl:Infrastructure (dbpedia-owl:Infrastructure is a subClass of dbpedia-owl:ArchitecturalStructure). However, these airports can

© Springer-Verlag Berlin Heidelberg 2016
N.T. Nguyen and R. Kowalczyk (Eds.): TCCI XXII, LNCS 9655, pp. 106–124, 2016.
DOI: 10.1007/978-3-662-49619-0_6

be described in other datasets as instances of dbpedia-owl:Building, which is a subClass of dbpedia-owl:ArchitecturalStructure. By this example, we shows that with common classes from an ontology (DBpedia[1] in this case), we could still encounter inconsistent problems while defining a new concept.

In our point of view, the inconsistent problem in which there are different versions of subject specification could be solved effectively using consensus theory [1].

In this paper, we present a new method based on the consensus theory to solving concept conflicts in ontology integration. Our approach is detailed and structured in this paper as follows: Sect. 2 shows some basic notions which are directly used in consensus-based knowledge integration. Section 3 formulates the problems of ontology integration and some drawbacks of current approaches for this problem. Next, we propose postulates and an algorithm for the integration problem. Section 4 presents ways to formulate distance functions which can be used in the most popular web ontology language, OWL 2. We define the distance functions between class expressions and data ranges. The paper concluded with discussions in Sect. 5.

2   Background

Consensus theory [1] is an appropriate tool to build collective intelligence. Several results of consensus theory using for knowledge integration have been proposed in [13]. In this section, we show some basic notions which are directly used in formalization of consensus-based model for knowledge integration as well as the problem of ontology integration.

2.1   Consensus Theory

By $\mathbf{U}$ we denote a finite set of objects representing possible values for a knowledge state. We also denote:

- $\prod_k(\mathbf{U})$ is the set of all k-element subsets (with repetitions) of set $\mathbf{U}$ ($k \in \mathbb{N}$, set of natural numbers).
- $\prod(\mathbf{U}) = \bigcup_{k \in \mathbb{N}} \prod_k(\mathbf{U})$ is the set of all nonempty subsets with repetitions of set $\mathbf{U}$. An element in $\prod(\mathbf{U})$ is called as a *conflict profile*.

Definition 1 - Distance function. *A distance function $d : \mathbf{U} \times \mathbf{U} \to [0,1]$ is defined so that it has these following features:*

1. Nonnegative: $\forall x, y \in \mathbf{U} : d(x,y) \geq 0$,
2. Reflexive: $\forall x, y \in \mathbf{U} : d(x,y) = 0 \Leftrightarrow x = y$,
3. Symmetrical: $\forall x, y \in \mathbf{U} : d(x,y) = d(y,x)$.

---

[1] http://datahub.io/dataset/dbpedia.

We call the space $(\mathbf{U}, d)$ which is defined in the above way as a distance space. With an $\mathsf{X} \in \prod(\mathbf{U})$, $M = |\mathsf{X}|$, $u \in \mathbf{U}$, we denote:

$$d(u, \mathsf{X}) = \sum_{x \in \mathsf{X}} d(u, x) \tag{1}$$

$$d_{t\_mean}(\mathsf{X}) = \frac{1}{M(M+1)} \sum_{x,y \in \mathsf{X}} d(x, y) \tag{2}$$

$$d_{min}(\mathsf{X}) = \frac{1}{M} . min_{u \in \mathbf{U}} \ d(u, \mathsf{X}) \tag{3}$$

Definition 2 - Consensus function. *By a consensus function in space $(\mathbf{U}, d)$, we mean a function*

$$C \colon \prod(\mathbf{U}) \to 2^{\mathbf{U}}.$$

For a conflict profile $\mathsf{X} \in \prod(\mathbf{U})$, the set $C(\mathsf{X})$ is called the representation of X, and an element in $C(\mathsf{X})$ is called a consensus of profile X. $C(\mathsf{X})$ is a normal set (without repetitions).

Consensus functions need to satisfy some postulates [13] in order to elect the "proper" representation(s) from a conflict profile. The mostly used consensus function are $\mathcal{O}_1$-functions. The functions $C(\mathsf{X}), \mathsf{X} \in \prod(\mathbf{U})$, of this kind satisfy the so-called $\mathcal{O}_1$-postulate [13]:

$$\big(x \in C(\mathsf{X})\big) \Rightarrow \big(d(x, \mathsf{X}) = min_{y \in \mathbf{U}} \ d(y, \mathsf{X})\big).$$

Definition 3 - Criteria for Consensus Susceptibility. *Not from any conflict profile we can choose a consensus solution in general and $\mathcal{O}_1$-consensus in specifically. We say that, profile X is susceptible to consensus in relation to postulate $\mathcal{O}_1$ iff:*

$$d_{t\_mean}(\mathsf{X}) \geq d_{min}(\mathsf{X}).$$

2.2   Consensus-Based Model for Knowledge Integration

Definition 4 - The $(\mathbf{A}, \mathbf{V})$ real world. *Let $\mathbf{A}$ is a set of attributes. Each attribute of $a \in \mathbf{A}$ has a set $\mathbf{V}_a$ of elementary values. We assume that a value of attribute a may be a subset of $\mathbf{V}_a$ as well as some element of $\mathbf{V}_a$. The set $2^{\mathbf{V}_a}$ is called as the super domain of attribute a. Letting*

$$\mathbf{V} = \bigcup_{a \in \mathbf{A}} \mathbf{V}_a,$$

the real world can be denoted by the pair $(\mathbf{A}, \mathbf{V})$.

For $\mathbf{T} \subseteq \mathbf{A}$. Let's denote

- $\mathbf{V_T} = \bigcup_{a \in \mathbf{T}} \mathbf{V}_a;$
- $\overline{2}^{\mathbf{V_T}} = \bigcup_{a \in \mathbf{T}} 2^{\mathbf{V}_a}.$

We have definitions of a *complex tuple* (or *tuple* for short) and an *elementary tuple* of type $\mathbf{T}$ as bellow.

Definition 5 - Complex tuple of type T. *We call a tuple of type* $\mathbf{T}$ *as a function*

$$r: \mathbf{T} \to \overline{2}^{\mathbf{V_T}}$$

such that $r(a) \subseteq \mathbf{V}_a$ *for all* $a \in \mathbf{T}$. *Instead of* $r(a)$ *we write* r_a *and a tuple of type* $\mathbf{T}$ *is written as* $r_\mathbf{T}$. *A tuple* $r_\mathbf{T}$ *may also be written as a set:*

$$r = \{(a, r_a) : a \in \mathbf{T}\}.$$

The set of all tuples of type $\mathbf{T}$ *is denoted by* $TUPLE(T)$.

Definition 6 - Elementary tuple of type T. *We call an elementary tuple of type* $\mathbf{T}$ *as a function*

$$r: \mathbf{T} \to \mathbf{V_T}$$

such that $r(a) \in \mathbf{V}_a$ *for all* $a \in \mathbf{T}$. *If* $\mathbf{V}_a = \emptyset$ *then* $r(a) = \varepsilon$, *where symbol* ε *represents a special value which is used when the domain is empty. The set of all elementary tuples of type* $\mathbf{T}$ *is denoted by* $ETUPLE(\mathbf{T})$.

Definition 7 - Sum of tuples. *The sum of two tuples* $r_\mathbf{T}$ *and* $r'_{\mathbf{T}'}$ *is a tuple* $r''_{\mathbf{T}''}$, *where* $\mathbf{T}'' = \mathbf{T} \cup \mathbf{T}'$ *and*

$$r''_a = \begin{cases} r_a \cup r'_a & \text{for } a \in \mathbf{T} \cap \mathbf{T}' \\ r_a & \text{for } a \in \mathbf{T} \backslash \mathbf{T}' \\ r'_a & \text{for } a \in \mathbf{T}' \backslash \mathbf{T} \end{cases}$$

Definition 8 - Product of tuples. *The product of two tuples* $r_\mathbf{T}$ *and* $r'_{\mathbf{T}'}$ *is a tuple* $r''_{\mathbf{T}''}$ *where* $\mathbf{T}'' = \mathbf{T} \cap \mathbf{T}'$ *and* $r''_a = r_a \cap r'_a$ *for each* $a \in \mathbf{T}''$.

Definition 9 - The $\prec$ relationship. *Let* $r \in TUPLE(\mathbf{T})$ *and* $r' \in TUPLE(\mathbf{T}')$ *where* $\mathbf{T} \subseteq \mathbf{T}'$. *We say that tuple* r *is included in tuple* r', *denoted as* $r \prec r'$, *if and only if* $r_a \subseteq r'_a$ *for each* $a \in \mathbf{T}$.

Definition 10 - The problem of knowledge integration. *Given a conflict profile* $\mathsf{X} = \{r_i \in TUPLE(\mathbf{T}_i) : \mathbf{T}_i \subseteq \mathbf{A} \text{ for } i = 1, 2, \ldots, n\}$, *one should determine a tuple* r^* *of type* $\mathbf{T}^* \subseteq \mathbf{A}$ *which best represents the given tuples.*

Since there is no assumption that the tuples in the conflict profile are the same type, the process of finding consensus tuple need to have a bit difficult steps rather than applying a consensus function in Definition 2. Nguyen [11] has defined six postulates for knowledge integration as below.

P_1. Closure of knowledge - 1
The type of the integration should be included in the sum of types of the profile elements; that is,

$$\mathbf{T}^* \subseteq \bigcup_{i=1}^{n} \mathbf{T}_i.$$

P_2. Closure of knowledge - 2
The integration should be included in the sum of profile elements; that is,

$$r^* \prec \bigcup_{i=1}^{n} r_i.$$

P_3. Consistency of knowledge
The common part of profile elements should be included in the integration; that is,

$$\bigcap_{i=1}^{n} r_i \prec r^*.$$

P_4. Superiority of knowledge - 1
For each attribute $a \in \mathbf{T}^*$, value $r^*(a)$ depends only on definite values from $\{r_i(a) : i = 1, 2, \ldots, n\}$.

P_5. Superiority of knowledge - 2
If sets of attributes $\mathbf{T}_i(i = 1, 2, \ldots, n)$ are disjoint with each other then

$$r^* = \left[\bigcup_{i=1}^{n} r_i \right]_{T^*}$$

where $\left[\bigcup_{i=1}^{n} r_i \right]_{T^*}$ is the sum $\bigcup_{i=1}^{n} r_i$ restricted to attributes from set $\mathbf{T}^*$.

P_6. Maximal similarity
Let d_a be a function measuring the distance between values of attribute $a \in \mathbf{A}$ then the difference between integration r^* and the profile elements should be minimal in the sense that for each $a \in \mathbf{T}^*$, the sum of distance

$$\sum_{r \in \mathbf{Z}_a} d\big(r^*(a), r(a)\big)$$

should be minimal, where,

$$\mathbf{Z}_a = \{r_i(a) : r_i(a) \text{ is } definite, \ i = 1, 2, \ldots, n\}.$$

3   Solving Conflicts in Ontology Integration

A definition for ontologies integration has proposed in [13]: *"For given ontologies $O_1, O_2, \ldots, O_n$ one should determine one ontology O^* which best represents them"*. The key problem is that we have to solve conflicts or inconsistencies between entities in the ontologies. We classify inconsistencies between ontologies into the following levels:

- Inconsistency on the instance level: There are several instances with the same name having different descriptions in different ontologies.
- Inconsistency on the concept level: There are several concepts with the same name having different structures in different ontologies.
- Inconsistency on the relation level: Between the same two concepts there are inconsistent relations in different ontologies.

In the last ten years, the problem of ontology integration has been a challenging issue which attracts several efforts of ontology research community. Most of approaches in ontology integration are based on process of determining mappings of entities in ontologies (ontology matching). String comparison on entities' names, or even on terms extracted from entities' descriptions (by using NLP techniques) can be used in this process [6]. Moreover, advanced techniques can be applied to improve result of ontology matching. Nikolov et al. [14] used instance-level coreference links to and from individuals defined in third-party repositories as background knowledge for schema-level ontology matching. Jain et al. [8,9] proposed systems (BLOOMS, BLOOMS+) which use information from Wikipedia category hierarchy (or existing upper-level ontologies) in process of finding alignments. However, after the matching process, what should we do with inconsistent concepts (which have the same name but different structures in ontologies)?

Recently, consensus theory [1] has been used for resolving conflicts in ontologies and gained several good results: In [13] (2007) and [4] (2011), authors proposed algorithms for integrating inconsistent ontologies on concept level. However, in our point of view, these algorithms only focus on attribute list of the integration structure of the concept. Our approach proposes an algorithm not only generate the attribute list of the integration structure of the concept but also calculate domains of attributes.

Definition 11 - Ontology. *An ontology is a quadruple* $\langle \mathbf{C}, \mathbf{I}, \mathbf{R}, \mathbf{Z} \rangle$, *where:*

- $\mathbf{C}$ *is a set of concepts (classes).*
- $\mathbf{I}$ *is a set of instances of concepts.*
- $\mathbf{R}$ *is a set of binary relations defined on* $\mathbf{C}$.
- $\mathbf{Z}$ *is a set of axioms which are formulas of first-order logic and can be interpreted as integrity constraints or relationships between instances and concepts, and which cannot be expressed by the relations in set* $\mathbf{R}$, *nor as relationships between relations included in* $\mathbf{R}$.

A domain ontology that refers to the real world $(\mathbf{A}, \mathbf{V})$ *is called* $(\mathbf{A}, \mathbf{V})$-*based.*

Definition 12 - Structure of a concept. *A concept in an* $(\mathbf{A}, \mathbf{V})$-*based ontology is defined as a triple* $(c, \mathbf{A}^c, \mathbf{V}^c)$, *where:*

- c *is the unique name of the concept,*
- $\mathbf{A}^c \subseteq \mathbf{A}$ *is a set of attributes describing the concept,*
- $\mathbf{V}^c = \bigcup_{a \in \mathbf{A}^c} \mathbf{V}_a$ *is the domain of attributes* $(\mathbf{V}^c \subseteq \mathbf{V})$.

The pair $(\mathbf{A}^c, \mathbf{V}^c)$ *is called the structure of concept* c.

Definition 13 - Relations between attributes. *Two attributes a, b in structure of a concept can have following relations:*

- *Equivalence: a is equivalence to b, denoted as $a \leftrightarrow b$, if a and b reflect the same feature for instances of the concept. For example, occipation $\leftrightarrow$ job.*
- *Generalization: a is more general than b, denoted as, $a \rightarrow b$, if information given by property a including information given by property b. For example: $dayOfBirth \rightarrow age$.*
- *Contradiction: a is contradictory with b, denoted as $a \downarrow b$, if their domains are the same two-element set and values of them for the same instance are contradictory. For example: $isFree \downarrow isLent$, where $\mathbf{V}_{isFree} = \mathbf{V}_{isLent} = \{true, false\}$ which can be used to describe instances in the Book concept whether its instances' property $isFree$ changed to $isLent$.*

Definition 14 - The ontology integration problem on the concept level.
Let $O_1, O_2, \ldots, O_n$, $(n \in \mathbb{N})$ are $(\mathbf{A}, \mathbf{V})$-based ontologies. Let the same concept c belong to O_i is $(c, \mathbf{A}^i, \mathbf{V}^i)$, $i = 1, 2, \ldots, n$. From the profile $\mathsf{X} = \{(\mathbf{A}^i, \mathbf{V}^i) : i = 1, 2, \ldots, n\}$, we have to determine the pair $(\mathbf{A}^*, \mathbf{V}^*)$ which presents the best structure for the concept c.

The ontology integration problem on the concept level of Definition 14 can be re-formulated to the knowledge integration of Definition 10 as we can see, a structure of the concept c in ontology O_i is an *elementary tuple* of type $\mathbf{A}^i$. With this starting point, we propose an algorithm trying to satisfy as much postulates as possible the six ones which are mentioned in Sect. 2.2.

3.1 Postulates for Determination of Pair $(\mathbf{A}^*, \mathbf{V}^*)$

Inspired by [13], we formulate the following postulates for determination of pair $(\mathbf{A}^*, \mathbf{V}^*)$.

R1. For $a, b \in \mathbf{A} = \bigcup_{i=1}^{n} \mathbf{A}^i$ and $a \leftrightarrow b$, all occurrences of a in all sets $\mathbf{A}^i$ may be replaced by attribute b or vice versa.

R2. If in any set $\mathbf{A}^i$ attributes a and b appear simultaneously and $a \rightarrow b$ then attribute b may be removed.

R3. For $a, b \in \mathbf{A} = \bigcup_{i=1}^{n} \mathbf{A}^i$ and $a \downarrow b$, all occurrences of a in all sets $\mathbf{A}^i$ may be replaced by attribute b or vice versa.

R4. Occurrence of an attribute in set $\mathbf{A}^*$ should be dependent only on the appearances of this attributes in sets $\mathbf{A}^i$.

R5. An attribute a appears in set $\mathbf{A}^*$ if it appears in at least half of sets $\mathbf{A}^i$.

R6. Set $\mathbf{A}^*$ is equal to $\mathbf{A}$ after applying postulates P1-P3.

R7. For each attribute $a \in \mathbf{A}^*$, its domain $\mathbf{V}_a^*$ is determined so that:

$$d_a(\mathbf{V}_a^*, \mathsf{X}_a) = min\{d_a(\mathbf{V}_a, \mathsf{X}_a) : \mathbf{V}_a \in \mathbf{U}_a\},$$

where:

- X_a is the conflict profile which is formulated from domains $V_a^i, i = 1, \ldots, n$.
- U_a is the universe set, contains all possible values for V_a.
- d_a is the distance function between elements in U_a.

Postulates R1–R6 are adapted to ones in [13]. We propose the R7 postulate to gain the result of consensus theory. More specifically, we use the $\mathcal{O}_1$ function to determine the optimal domain for the attribute $a \in A^*$. It is important to formulate distance space (U_a, d_a) for using $\mathcal{O}_1$ function to find the consensus domain. The important issue is about appropriately defining space distance (U_a, d_a) to compute the optimised solution for the ranges of properties in integration set.

With these postulates, we propose the integration algorithm, Algorithm 1 for conflict ontologies as below section.

3.2   Algorithm for Determining the Optimal Integration Based on the Consensus

Based on postulates that are presented in previous section, we propose an algorithm for determining integration structure for concept c from element ontologies $O_1, O_2, \ldots, O_n$ (Algorithm 1).

As [13] pointed, not for each conflict profile of tuples, we can find a consensus which satisfies all of postulates. But, we will show that, the algorithm satisfies postulates P_1, P_2, P_3, P_4, P_5 and partly satisfies P_6 postulate as the following theorem.

Theorem 1. *Algorithm 1 has the following properties for any profile X:*

(a) The consensus determined by Algorithm 1 satisfies postulates P_1, P_2, P_3, P_4.
(b) If we do not apply postulate R_5 in the algorithm, the consensus determined by Algorithm 1 satisfies postulate P_5.
(c) The consensus determined by Algorithm 1 partly satisfies postulate P_6.

Proof. (a) We show the algorithm satisfies each of postulate in $P1 - P4$.

- At **Step 1** of the algorithm, we start out initial set of A^* as

$$A^* \subseteq \bigcup_{i=1}^{n} A^i.$$

  After that, we only remove elements from A^*. So, the algorithm satisfies postulate P_1 (closure of knowledge - 1).
- By setting $U_a = \bigcup_{i=1}^{n} V_a^i$ as input of the algorithm, we can make sure that

$$V_a^* \in \bigcup_{i=1}^{n} V_a^i$$

  This means, the algorithm satisfies postulate P_2 (closure of knowledge - 2).

Algorithm 1. Determine the optimal integration structure for concept

Input:
- Conflict profile $X = \{(\mathbf{A}^i, \mathbf{V}^i), i = 1, \ldots, n\}$, where $(\mathbf{A}^i, \mathbf{V}^i)$ is structure of concept c in ontology O_i.
- $\mathbf{U}_a$ is the universe set, contains all possible values for $\mathbf{V}_a$.
- d_a is the distance function between elements in $\mathbf{U}_a$.

Output: Pair $(\mathbf{A}^*, \mathbf{V}^*)$ present the best structure of concept c.

begin

Step 1   Set $\mathbf{A}^* := \bigcup_{i=1}^n \mathbf{A}^i$;

Step 2   **foreach** $a, b \in \mathbf{A}^*$ **do**

    if *((a ↔ b) and (a does not occur in relationships with other attributes from $\mathbf{A}^*$))* **then**

       | Set $\mathbf{A}^* := \mathbf{A}^* \backslash \{a\}$;

    end

    if *((a → b) and (b does not occur in relationships with other attributes from $\mathbf{A}^*$))* **then**

       | Set $\mathbf{A}^* := \mathbf{A}^* \backslash \{b\}$;

    end

    if *((a ↓ b) and (b does not occur in relationships with other attributes from $\mathbf{A}^*$))* **then**

       | Set $\mathbf{A}^* := \mathbf{A}^* \backslash \{b\}$;

    end

end

Step 3   **foreach** $a \in \mathbf{A}^*$ **do**

    if *(the number of occurrences of a in pairs $(\mathbf{A}^i, \mathbf{V}^i)$ is smaller than $\frac{n}{2}$)* **then**

       | Set $\mathbf{A}^* := \mathbf{A}^* \backslash \{a\}$;

    else

       Set $X_a := \{\mathbf{V}_1, \mathbf{V}_2, \ldots, \mathbf{V}_k\}$ where V_j is the domain of attribute a in pair $(\mathbf{A}^i, \mathbf{V}^i)$, $j = 1, \ldots, k$ and $i = 1, \ldots, n$;

       if *(X_a is susceptible to consensus in relation to postulate $\mathcal{O}_1$)* **then**

          | Determine $\mathbf{V}_a^*$ as $\mathcal{O}_1$ consensus in distance space $(\mathbf{U}_a, d_a)$:
            $d(\mathbf{V}_a^*, X_a) = min\{d(\mathbf{V}_a, X_a) : \mathbf{V}_a \in \mathbf{U}_a\}$;

          | Set $\mathbf{V}_a^*$ as domain of attribute a in $\mathbf{A}^*$;

       else

          | Set $\mathbf{A}^* := \mathbf{A}^* \backslash \{a\}$;

       end

    end

end

Step 4   **foreach** $a \in \mathbf{A}^*$ **do**

    if *(there is a relationship a ↔ b or a → b or a ↓ b)* **then**

       | $\mathbf{A}^* := \mathbf{A}^* \cup \{b\}$

    end

end

end

– Next, we prove that the algorithm satisfies postulate P_3.
Let us assume that, $(\overline{a}, \overline{\mathbf{V}}) \in (\mathbf{A^i}, \mathbf{V^i})$ for $i = 1, \ldots, n$. In this case, the number of attribute $\overline{a}$ in the profile equals to n $(\geq \frac{n}{2})$. So, as in **Step 3** of the algorithm, we formulate the conflict profile $\mathsf{X}_{\overline{a}} = \{n * \overline{\mathbf{V}}\}$ (n elements of $\overline{\mathbf{V}}$).
We have

$$d_{t\_mean}(\mathsf{X}_{\overline{a}}) = \frac{1}{n.(n+1)} \cdot \sum_{x,y \in \mathsf{X}_{\overline{a}}} d_{\overline{a}}(x, y)$$

$$= 0 \quad (d_{\overline{a}}(x, y) = 0 \; for \; all \; x, y \in \mathsf{X}_{\overline{a}});$$

We also have,

$$d_{\overline{a}}(x, \mathsf{X}_{\overline{a}}) \geq 0 \;\; \forall x \in \mathbf{U}_{\overline{a}}.$$
$$d_{\overline{a}}(\overline{\mathbf{V}}, \mathsf{X}_{\overline{a}}) = 0.$$

So, $d_{min}(\mathsf{X}_{\overline{a}}) = \frac{1}{n}.min\{d_{\overline{a}}(x, \mathsf{X}_{\overline{a}}) \mid x \in \mathbf{U}_{\overline{a}}\} = 0$.
Thus, we have:

- $d_{t\_mean}(\mathsf{X}_{\overline{a}}) \geq d_{min}(\mathsf{X}_{\overline{a}})$. This means, $\mathsf{X}_{\overline{a}}$ is susceptible to consensus in relation to postulate $\mathcal{O}_1$; and
- $\mathbf{V}^* = \overline{\mathbf{V}}$.

Finally, we have:

$$(\overline{a}, \overline{\mathbf{V}}) \in (\mathbf{A^i}, \mathbf{V^i}) \Rightarrow (\overline{a}, \overline{\mathbf{V}}) \in (\mathbf{A^*}, \mathbf{V^*}) \quad \text{for } i = 1, \ldots, n$$

This means, the algorithm satisfies postulate P_3 (Consistency of knowledge).
– For each attribute $a \in \mathbf{A}^*$, as in the algorithm, $\mathbf{V}_a^*$ is determined only in the set $\mathbf{U}_a$. So, the postulate P_4 (Superiority of knowledge - 1) is satisfied.

(b) Let us assume that, sets of $\mathbf{A}^i$ $(i = 1, \ldots, n)$ are disjoint with each other and we do not use the R_5 postulate in the algorithm. It means that, we do not remove any attribute in the initial set $\mathbf{A}^* = \bigcup_{i=1}^{i=n} \mathbf{A}^i$.
And, as in the algorithm, for each attribute $a \in \mathbf{A}^*$, we formulate the profile X_a which contains only one element. In this case, we can clearly see that:
– The profile is susceptible to consensus in relation to postulate $\mathcal{O}_1$; and,
– The unique element in the profile is also the consensus.
This means, the algorithm satisfies postulates P_5 (Superiority of knowledge - 2).

(c) As in the algorithm, for each attribute $a \in \mathbf{A}^*$, we determine the $\mathbf{V}_a^*$ by choosing elements in the set $\mathbf{U}_a$, which is formulate at input of algorithm as set of values $\mathbf{V}_a^i$ $(i = 1, \ldots, n)$. Thus, we say the algorithm partly satisfies the postulate P_6 (Maximal similarity).

$\square$

We have remarks for this algorithm:

- The complexity of the algorithm is $O(n.m^2)$ where $m = \left|\bigcup_{i=1}^{n} \mathbf{A}^i\right|$ (n is the number of elements in the profile X, m is the number of different attributes from sets $\mathbf{A}^i$, $i = 1, \ldots, n$).
- The algorithm determines consensus structure for concept c in both component: attributes and their domains.
- The decision to apply the postulate R_5 or not is based on answer for the question "Does we want to include all attributes of the concept which are asserted in participant ontologies?"
    - The "Yes" answer reflects that, we want to exploit knowledge of all participants. It useful to know that, in the Open Assumption World, we should collect as much facts as possible for the knowledge base!
    - The "No" answer reflects that, we want to have a high agreement of participants.
- We can make modification in **Step 3** of the algorithm to get some interesting results: If X_a is not susceptible in relation to $\mathcal{O}_1$, we can choose $\mathbf{V}_a^*$ as sum of its domains in pairs $(\mathbf{A}^i, \mathbf{V}^i)$.

4   Formulating Distance Functions

In order to apply the Algorithm 1, we have to formulate distance d_a between two domains of the attribute. In ontologies, there are two kinds of properties: *DataType Properties* link individuals to data values (liteals) and *Object Properties* link individuals to individuals. The following section shows different ways to formulate (1) distance between two Class Expressions (used for describing values or ranges of Object Properties), and (2) distance between two Data Ranges (used for describing values or ranges of Data Properties).

4.1   Formulating Distance Between Two ClassExpressions

In the structural specification of OWL 2,[2] class expressions are represented by `ClassExpression`. Classes are the simplest form of class expressions. The other, complex, class expressions, can be constructed as below:

```
ClassExpression :=
    Class |
    ObjectIntersectionOf | ObjectUnionOf | ObjectComplementOf | ObjectOneOf |
    ObjectSomeValuesFrom | ObjectAllValuesFrom | ObjectHasValue | ObjectHasSelf |
    ObjectMinCardinality | ObjectMaxCardinality | ObjectExactCardinality |
    DataSomeValuesFrom | DataAllValuesFrom | DataHasValue |
    DataMinCardinality | DataMaxCardinality | DataExactCardinality
```

We define a semantic distance between two concepts in a so-called referenced ontology. Then we formulate the distance between two Class Expressions based on that semantic distance.

---

[2] http://www.w3.org/TR/owl2-syntax.

There are several ways to measure the similarity between concepts in an ontology. In this paper, we use idea of Jike Ge and Yuhui Qiu [7]. According to authors, we allocate weight values to the edges between concepts:

$$w(parent, child) = 1 + \frac{1}{2^{depth(child)}}$$

where, $depth(child)$ presents the depth of concept $child$ from the root concept in ontology hierarchy. The semantic distance between concepts can be determined using Algorithm 2 [7].

Algorithm 2. Calculate semantic distance between concepts

Input: Concepts c_1, c_2 in ontology hierarchy.
Output: Semantic distance between c_1, c_2, denoted as $Sem\_Dis(c_1, c_2)$
begin
    if *(c_1, c_2 is the same concept)* **then**
        $Sem\_Dis(c_1, c_2) := 0$;
    else if *(there exists the direct path relation between c_1 and c_2)* **then**
        $Sem\_Dis(c_1, c_2) := w(c_1, c_2)$;
    else if *(there exists the indirect path relations between c_1 and c_2)* **then**
        Determine shortestPath(c_1, c_2) is the shortest path from c_1 to c_2 in the ontology hierarchy;
        $Sem\_Dis(c_1, c_2) := \sum_{(c_i, c_j) \in shortestPath(c_1, c_2)} w(c_i, c_j)$;
    else
        Determine $cpp = $ the nearest common parent concept of the two concepts c_1, c_2;
        $Sem\_Dis(c_1, c_2) := min\{Sem\_Dis(c_1, cpp)\} + min\{Sem\_Dis(c_2, cpp)\}$;
    end
end

We can see clearly that, the Sem_Dis function is not normalized, i.e. its values may be out of $[0, 1]$. We can normalise it like this:

$$d: \mathbf{U} \times \mathbf{U} \rightarrow [0, 1]$$

$$d(c_1, c_2) \mapsto 1 - \frac{1}{Sem\_Dis(c_1, c_2) + 1}$$

where $\mathbf{U}$ is the set of named concepts in the referenced ontology.

Next, we will formulate the distance between two Class Expressions by repeatedly applying following rules:

(i) $d(ObjectIntersectionOf(CE_1, \ldots, CE_n), CE) = min\{d(CE_1, CE), \ldots, d(CE_n, CE)\}$

(ii) $d(ObjectUnionOf(CE_1, \ldots, CE_n), CE) = \frac{1}{n}.(d(CE_1, CE) + \cdots + d(CE_n, CE))$

(iii) In the case of ObjectOneOf, assume that $CE = ObjectOneOf$ $(obj_1, \ldots, obj_n)$ and obj_i is an instance of concept CE_i, $i = 1, \ldots, n$. We can use the class expression below to calculate the distance instead of CE: $ObjectUnionOf(CE_1, \ldots, CE_n)$.

It means,
$$d\big(ObjectOneOf(obj_1, \ldots, obj_n), CE\big) = d\big(ObjectUnionOf(CE_1, \ldots, CE_n), CE\big)$$

(iv) In the case of ObjectComplementOf, we assume that $CE_1 = ObjectComplementOf(CE_2)$. If CE_1 is not existed in the set of named concepts of the referenced ontology, we can approximately calculate the distance as
$$d(CE, ObjectComplementOf(CE_2)) = 1 - d(CE, CE_2).$$

 (v) In the remaining cases, we can approximately calculate distance by simply using domain of the object property. With assuming that the Object Property Expression OPE has domain as CE', we have:
 - $d\big(ObjectSomeValuesFrom(OPE, CE), CE\big) = d(CE', CE)$
 - $d\big(ObjectAllValuesFrom(OPE, CE), CE\big) = d(CE', CE)$
 - $d\big(ObjectMinCardinality(n, OPE, CE), CE\big) = d(CE', CE)$
 - $d\big(ObjectMaxCardinality(n, OPE, CE), CE\big) = d(CE', CE)$
 - $d\big(ObjectExactCardinality(n, OPE, CE), CE\big) = d(CE', CE)$
 - $d\big(ObjectHasValue(OPE, a), CE\big) = d(CE', CE)$
 - $d\big(ObjectHasSelf(OPE), CE\big) = d(CE', CE)$
 - $d\big(DataSomeValuesFrom(OPE, CE), CE\big) = d(CE', CE)$
 - $d\big(DataAllValuesFrom(OPE, CE), CE\big) = d(CE', CE)$
 - $d\big(DataMinCardinality(n, OPE, CE), CE\big) = d(CE', CE)$
 - $d\big(DataMaxCardinality(n, OPE, CE), CE\big) = d(CE', CE)$
 - $d\big(DataExactCardinality(n, OPE, CE), CE\big) = d(CE', CE)$
 - $d\big(DataHasValue(OPE, DR), CE\big) = d(CE', CE)$.

4.2   Formulating Distance Between Two Data Ranges

Data ranges can be used in restrictions on data properties. The syntax to formulate a data range in OWL 2 is:

```
DataRange :=
    Datatype |
    DataIntersectionOf |
    DataUnionOf |
    DataComplementOf |
    DataOneOf |
    DatatypeRestriction
```

The simplest data ranges are datatypes. The DataIntersectionOf, DataUnionOf, and DataComplementOf data ranges provide for the standard set-theoretic operations on data ranges. The DataOneOf data range consists of exactly the specified set of literals. Finally, the DatatypeRestriction data range restricts the value space of a datatype by a constraining facet.

Each datatype is identified by an IRI and is defined by the following components:

- The value space is the set of values of the datatype. Elements of the value space are called data values.
- The lexical space is a set of strings that can be used to refer to data values. Each member of the lexical space is called a lexical form, and it is mapped to a particular data value.
- The facet space is a set of pairs of the form (F, v) where F is an IRI called a constraining facet, and v is an arbitrary data value called the constraining value. Each such pair is mapped to a subset of the value space of the datatype.

In the context of measuring distance, we can omit the lexical space component of a Datatype. We can also classify Datatypes to 3 categories:

(1) Datatypes which have value space as a set of numbers, such as `owl:real`, `owl:rational`, `owl:decimal`, `xsd:double`, `xsd:float`, `xsd:integer`, `xsd:int`, `xsd:long`, `xsd:short`, ...;
(2) Datatypes which have value space can be directly mapped to set of numbers, such as `xsd:boolean`, `xsd:dateTime`;
(3) "Non-numeric" Datatypes, such as `xsd:string`, `xsd:hexBinary`, ...

In this paper, we only discuss ways to calculate distance of "numeric" Datatypes. The facet space component of these Datatypes is used to formulate an interval of number.

Example 1. We can present an interval $[12, 20]$ in OWL 2 as

```
DatatypeRestriction(xsd:integer
    xsd:maxInclusive"20"^^xsd:integer
    xsd:minInclusive "12"^^xsd:integer)
```

Next, we formulate distance between intervals. For this purpose, we need some definitions.

We present an interval by a pair $[i_*, i^*]$, where i_* is the beginning of the interval and i^* is the ending of the interval, $i_* \leq i^*$. Intervals can not only present range of numbers but also present well a single number. For example, $[12, 20]$ represents an interval with scope from 12 to 20, and $[12, 12]$ represents a single value 12. There are several ways to calculate distance between intervals, but in this paper, we will use the following distance, which was used in [12, 15].

Definition 15 - Distance between two intervals. *Distance δ_R between two intervals r and q equals the sum of:*

- *half of the length of the part of r which is outside of q,*
- *half of the length of the part of q which is outside of r,*
- *the length of span between r and q.*

Function δ_R suits very well for calculating difference between intervals and single values. Zgrzywa [15] also shown that, δ_R can be calculated in an very easy and convenient manner.

Theorem 2 *[15]. Let set* **V** *contain elements which can be ordered. Let the elements from set* **V** *fulfill the equation:* $|v_i - v_j| + |v_j - v_k| = |v_i - v_k|$, *when* $v_i \le v_j \le v_k$. *Let* $r = [r_*, r^*]$ *and* $q = [q_*, q^*]$ *be intervals of elements from set* **V**. *The distance* δ_R *between intervals* r *and* q *equals:*

$$\delta_R = \frac{|q^* - r^*| + |q_* - r_*|}{2}.$$

Example 2. By using Theorem 2, we can calculate:

- $\delta_R([12, 20], [16, 29]) = \frac{|20-29|+|12-16|}{2} = 6$.;
- $\delta_R([12, 12], [20, 20]) = 8$.

As forementioned, a DataRange can be formulate using others "basic" DataRanges:

- `Datatype`
- `DataIntersectionOf`
- `DataUnionOf` D
- `DataComplementOf`
- `DataOneOf`

So, using the distance function δ_R, we can calculate distance between two numeric DataRange by repeatedly applying following rules:

(i) In the case of `DataIntersectionOf`, we determine the intersection of DataRange before calculating the final distance. It means:

$$\delta_R(DR, DataIntersectionOf(DR_1, \ldots, DR_n)) = \delta_R(DR, DR')$$

where $DR' = DataIntersectionOf(DR_1, \ldots, DR_n)$.

(ii) In the case of `DataComplementOf`, we determine the complement of DataRange before calculating the final distance. It means:

$$\delta_R(DR, DataComplementOf(DR_1)) = \delta_R(DR, DR')$$

where $DR' = DataComplementOf(DR_1)$.

(iii) With `DataUnionOf`, we use the following equation:

$$\delta_R(DR, DataUnionOf(DR_1 \ldots DR_n)) = \frac{\delta_R(DR, DR_1) + \cdots + \delta_R(DR, DR_n)}{n};$$

(iv) In the case of `DataOneOf`, we can use the same way as in the case of `DataUnionOf` (Because a single value can be treated as an interval).

Example 3. Calculate distance between interval $[12, 20]$ and $DataUnionOf$ $([1, 13], [15, 30])$:

$$\delta_R([12, 20], DataUnionOf([1, 13], [15, 30])) = \frac{\delta_R([12, 20], [1, 13]) + \delta_R([12, 20], [15, 30])}{2}$$

$$= \frac{9 + 6.5}{2}$$

$$= 7.75.$$

Note that, the distance function δ_R is not normalized, i.e. its values may be out of $[0, 1]$. We can choose a number N so that the following function d_R is normalized:

$$d_R : \text{DataRange} \times \text{DataRange} \rightarrow [0, 1]$$

$$d_R(DR_1, DR_2) \mapsto \frac{\delta_R(DR_1, DR_2)}{N}$$

4.3   An Example for the Algorithm

We consider a small example for our algorithm: Let $(\mathbf{A}, \mathbf{V})$ is a real world where:

- $\mathbf{A} = \{cid, isTaughtBy, isFinish, isActive, sched, tkb\}$[3]
- $\mathbf{V}_{cid} = [1, 1000]$
- $\mathbf{V}_{isTaughtBy} = \{AscProf, Prof, AssiProf, AcademicStaffMember\}$
- $\mathbf{V}_{isFinish} = \{Yes, No\}$
- $\mathbf{V}_{isActive} = \{Yes, No\}$
- $\mathbf{V}_{sched} = \{Mon, Tue, Wed, Thurs, Fri, Sat, Sun\}$
- $\mathbf{V}_{tkb} = \{2, 3, 4, 5, 6, 7, 8\}$.

Relationships between the attributes: $\{tkb \leftrightarrow sched, isFinish \downarrow isActive\}$
Concepts of ontologies reference to ontology $O_{REF-TREE}$ (Fig. 1).

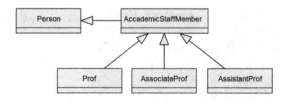

Fig. 1. Excerpt of referenced ontology $O_{REF-TREE}$

First, we calculate weight of edges in ontology $O_{REF-TREE}$:

- $w[Person, AcademicSM] = 1 + 1/2 = 1.5$
- $w[AcademicSM, AscProf] = 1 + 1/2^2 = 1.25$

---

[3] *tkb* is an acronym for *"thoi khoa bieu"* in Vietnamese, which is equals to *"schedule"* in English.

– $w[AcademicSM, Prof] = 1 + 1/2^2 = 1.25$
– $w[AcademicSM, AssiProf] = 1 + 1/2^2 = 1.25.$

Table 1. Structures of concept $Course$ from 5 ontologies

Ontology	Structure of concept $Course$
O_1	$\{(cid, [1, 1000]), (isActive, V_{isActive}), (sched, V_{sched}), (isTaughtBy, AssiProf)\}$
O_2	$\{(cid, [1, 1000]), (isFinish, V_{isFinish}), (isTaughtBy, ObjectUnionOf(AscProf, Prof))\}$
O_3	$\{(isActive, V_{isActive}), (tkb, V_{isFinish}), (cid, [1, 1000])\}$
O_4	$\{(cid, [1, 1000]), (isTaughtBy, ObjectUnionOf(AscProf, Prof))\}$
O_5	$\{(cid, [1, 200]), (isTaughtBy, AscProf)\}$

As the information provided in Table 1, step by step, we have results adapted to the algorithm:

– Step 1: $\mathbf{A}^* = \{cid, isActive, sched, isTaughtBy, isFinish, tkb\}$.
– Step 2: Remove 2 attributes $isFinish$ and tkb from $\mathbf{A}^*$. After this step, we have: $\mathbf{A}^* = \{cid, isActive, sched, isTaughtBy\}$.
– Step 3:
  - Consider cid: This attribute occurs 4 times in sets $\mathbf{A}^i$. Set $\mathsf{X}_{cid} = \{3 * [1, 1000], [1, 200]\}$. We can determine consensus of profile X_{cid} is $[1, 1000]$. So we have $\mathbf{V}^*_{cid} = [1, 1000]$.
  - Similarly, consider $isActive$: This attribute occurs 3 times ($> 5/2$), and its domain is $V^*_{isActive} = \{Yes, No\}$.
  - Consider $sched$: It occurs 2 times ($< 5/2$). So we remove it from $\mathbf{A}^*$. Now, we have $\mathbf{A}^* = \{cid, isActive, isTaughtBy\}$.
  - Consider $isTaughtBy$. It occurs 3 times ($> 5/2$) in sets $\mathbf{A}^i$. Set $\mathsf{X}_{isTaughtBy} = \{2 * ObjectUnionOf(AscProf, Prof), AscProf, AssiProf\}$. We easily get these following results:
    * $d(Prof, AssiProf) = d(Prof, AscProf) = d(AssiProf, AscProf) = 0.71$
    * $d(ObjectUnionOf(AscProf, Prof), \mathsf{X}_{isTaughtBy}) = \frac{3}{2} \times 0.71.$
    * $d(AscProf, \mathsf{X}_{isTaughtBy}) = 2 \times 0.71.$
    * $d(AssiProf, \mathsf{X}_{isTaughtBy}) = 3 \times 0.71.$
    * $d_{min}(\mathsf{X}_{isTaughtBy}) = \frac{1}{4} \times d(ObjectUnionOf(AscProf, Prof), \mathsf{X}_{isTaughtBy}) = \frac{3}{8} \times 0.71.$
    * $d_{t\_mean}(\mathsf{X}_{isTaughtBy}) = \frac{1}{4 \times 5}(8 \times 0.71) = \frac{2}{5} \times 0.71.$
    We have $d_{t\_mean}(\mathsf{X}_{isTaughtBy}) \geq d_{min}(\mathsf{X}_{isTaughtBy})$. So profile $\mathsf{X}_{isTaughtBy}$ is susceptible in relation to postulate $\mathcal{O}_1$. And the domain of $isTaughtBy$ is $V^*_{isTaughtBy} = ObjectUnionOf(AscProf, Prof)$.
– Step 4: Add the attribute $isFinish$ back to $\mathbf{A}^*$.
Finally, we have the structure of the concept $course$ as followed:

$$(\mathbf{A}^*, \mathbf{V}^*) = \{(cid, [1, 1000]), (isActive, \{Yes, No\}), (isFinish, \{Yes, No\}),$$
$$(isTaughtBy, ObjectUnionOf(AscProf, Prof))\}$$

5 Conclusion

In this paper, we present an algorithm for solving conflict on concept level by finding the consensus for ontology integration. The algorithm satisfies most of postulates which are proposed in consensus-based knowledge integration model [13]. We also show different ways to formulate distance functions between attributes' values. The paper also confirms that consensus theory is an appropriate and effective way for ontology integration problem.

As the future work, we would like to analyse the opportunities of using other consensus functions for determining consensus integration. We also would like to apply the approach of this paper for other level of conflict in ontologies: The distance functions between two intervals in Sect. 4.2 can be applied in consensus-based model for knowledge integration to resolve instance-level conflicts in ontology integration [13], which can be also applied in the area of data fusion for Linked Data applications [2].

References

1. Barthélemy, J.P., Janowitz, M.F.: A formal theory of consensus. SIAM J. Discrete Math. **4**(3), 305–322 (1991)
2. Bizer, C., Heath, T., Berners-Lee, T.: Linked data-the story so far. Int. J. Semant. Web Inf. Syst. (IJSWIS) **5**(3), 1–22 (2009)
3. Dastgheib, S., Mesbah, A., Kochut, K.: MOntage: building domain ontologies from linked open data. In: Proceedings of the 2013 IEEE Seventh International Conference on Semantic Computing, ICSC 2013, pp. 70–77. IEEE Computer Society, Washington, DC (2013). http://dx.org/10.1109/ICSC.2013.21
4. Duong, T.H., Nguyen, N.T., Kozierkiewicz-Hetmanska, A., Jo, G.: Fuzzy ontology integration using consensus to solve conflicts on concept level. In: ACIIDS Posters, pp. 33–42 (2011)
5. Euzenat, J.: Towards a principled approach to semantic interoperability. In: Proceedings of the 2001 Workshop on Ontology and Information Sharing, pp. 19–25 (2001)
6. Euzenat, J., Shvaiko, P.: Ontology Matching, 2nd edn. Springer, Heidelberg (2013)
7. Ge, J., Qiu, Y.: Concept similarity matching based on semantic distance. In: Proceedings of the Fourth International Conference on Semantics Knowledge and Grid, pp. 380–383. IEEE (2008)
8. Jain, P., Hitzler, P., Sheth, A.P., Verma, K., Yeh, P.Z.: Ontology alignment for linked open data. In: Patel-Schneider, P.F., Pan, Y., Hitzler, P., Mika, P., Zhang, L., Pan, J.Z., Horrocks, I., Glimm, B. (eds.) ISWC 2010, Part I. LNCS, vol. 6496, pp. 402–417. Springer, Heidelberg (2010). http://dl.acm.org/citation.cfm?id=1940281.1940308
9. Jain, P., Yeh, P.Z., Verma, K., Vasquez, R.G., Damova, M., Hitzler, P., Sheth, A.P.: Contextual ontology alignment of LOD with an upper ontology: a case study with proton. In: Antoniou, G., Grobelnik, M., Simperl, E., Parsia, B., Plexousakis, D., De Leenheer, P., Pan, J. (eds.) ESWC 2011, Part I. LNCS, vol. 6643, pp. 80–92. Springer, Heidelberg (2011). http://link.springer.com/10.1007/978-3-642-21034-1_6

10. Klein, M.: Combining and relating ontologies: an analysis of problems and solutions. In: Proceedings of the 2001 Workshop on Ontologies and Information Sharing, vol. 47, pp. 53–62 (2001). http://citeseerx.ist.psu.edu/viewdoc/summary?doi=10.1.1.13.3504

11. Nguyen, N.T.: Representation choice methods as the tool for solving uncertainty in distributed temporal database systems with indeterminate valid time. In: Monostori, L., Váncza, J., Ali, M. (eds.) IEA/AIE 2001. LNCS (LNAI), vol. 2070, pp. 445–454. Springer, Heidelberg (2001)

12. Nguyen, N.T.: Consensus system for solving conflicts in distributed systems. Inf. Sci. **147**(1), 91–122 (2002)

13. Nguyen, N.T.: Advanced Methods for Inconsistent Knowledge Management. Springer, London (2007)

14. Nikolov, A., Uren, V., Motta, E., de Roeck, A.: Overcoming schema heterogeneity between linked semantic repositories to improve coreference resolution. In: Gómez-Pérez, A., Yu, Y., Ding, Y. (eds.) ASWC 2009. LNCS, vol. 5926, pp. 332–346. Springer, Heidelberg (2009)

15. Zgrzywa, M.: Consensus determining with dependencies of attributes with interval values. J. Univ. Comput. Sci. **13**(2), 329–344 (2007)

Enhancing Collaborative Filtering Using Implicit Relations in Data

Manuel Pozo[1]([✉]), Raja Chiky[1], and Elisabeth Métais[2]

[1] Institut Supérieur d'Eléctronique de Paris, LISITE Lab,
28, Rue Notre-Dame-des-Champs, 75006 Paris, France
{manuel.pozo,raja.chiky}@isep.fr
[2] Conservatoire National des Arts Et Métiers, CEDRIC Lab, Paris, France
elisabeth.metais@cnam.fr
http://cedric.cnam.fr
http://lisite.isep.fr

Abstract. This work presents a Recommender System (RS) that relies on distributed recommendation techniques and implicit relations in data. In order to simplify the experience of users, recommender systems pre-select and filter information in which they may be interested in. Users express their interests in items by giving their opinion (explicit data) and navigating through the web-page (implicit data). The Matrix Factorization (MF) recommendation technique analyze this feedback, but it does not take more heterogeneous data into account. In order to improve recommendations, the description of items can be used to increase the relations among data. Our proposal extends MF techniques by adding implicit relations in an independent layer. Indeed, using past preferences, we deeply analyze the implicit interest of users in the attributes of items. By using this, we transform ratings and predictions into "semantic values", where the term semantic indicates the expansion in the meaning of ratings. The experimentation phase uses MovieLens and IMDb database. We compare our work against a simple Matrix Factorization technique. Results show accurate personalized recommendations. At least but not at last, both recommendation analysis and semantic analysis can be parallelized, alleviating time processing in large amount of data.

Keywords: Collaborative filtering · Distributed systems · Recommender system · Implicit interest

1 Introduction

The amount of information in the web has greatly increased in the past decade, and it is continuously growing. This makes tough the task of seeking information, and thus users of the Internet may feel overwhelmed when they do not find what they are looking for. These phenomenons has encouraged the development of Recommender Systems (RS). The aim of these systems is to pre-select and to filter information in webs in order to present first those in which users may be

© Springer-Verlag Berlin Heidelberg 2016
N.T. Nguyen and R. Kowalczyk (Eds.): TCCI XXII, LNCS 9655, pp. 125–146, 2016.
DOI: 10.1007/978-3-662-49619-0_7

more interested. This field has specially raised the attention of the e-commerce to offer personalized products (a.k.a. items) to users. Thus, one may observe these systems in movie platforms and online-shops, such as video media in Netflix or products in Amazon, but also in article researching and social networks, as Mendely, Google, Facebook or Twitter.

Typically, users express their interest in items by giving opinions (i.e. explicit data) and navigating through the web-pages (i.e. implicit data). For instance, users may rate items (e.g. movies) using a 0–5 stars scale, or they might just click on items links. This data is the interaction between users and items, and for the recommender it represents a feedback of users interest. Hence, recommender systems exploit this available information to predict future interests of users.

In literature, recommendation techniques may be classified in Content-Based (CB), Collaborative Filtering (CF) and Hybrid methods [1]. Content-based takes into account the domain of the recommendation (e.g. movies or books) and it recommends similar items to those the user liked in the past. This carries out over-specialization in recommendations and an item-domain dependency, e.g. always the same genre of movies. Collaborative filtering groups users according to their preferences or tastes, then it recommends items that people from the same group have already liked in the past [2]. Yet, it suffers from cold-start: the system have not yet information about new users/items in order to correctly group them [3]. Among these techniques, Matrix Factorization (MF) has demonstrated high accuracy and easy implementation [4]. In addition, it alleviates time processing in large amount of data by using a parallelizable algorithm. Hybrid methods combine different techniques to alleviate disadvantages and improve the general performance of the global system. In order to increase the quality of the recommendation, trend hybrid techniques seek more relations between users and items by implementing Semantic Technologies [5]. This enhances data representation and help to find out the reasons for which users may or may not be interested in a particular item. However, hybrid systems add complexity and item-domain dependency. In addition, the parallelization of the recommender becomes more difficult.

In this paper, we want to highlight a lack of knowledge in feedback: the interest of users in the attributes of items is hardly captured. Indeed, items contain many attributes (a.k.a. features, such as a movie genre or a movie actor), and moreover they may take several values (such a comedy genre or a concrete actor). This quantity of information makes very difficult to find out the interest of users in these aspects. In fact, (1) users are not willing to give too much explicit information about the features of items, and (2) the large amount of features makes explicit feedback in features inappropriate. For instance, users hardly would rate every actor in a movie.

We claim that the interest of users in these features may render predicted items more acceptable by users. We present a flexible and generic collaborative filtering system that relies on matrix factorization and implicit relations in data. We exploit the description of items and attributes to allow making implicit relations among data. This may help to discover the implicit interest of the user in the attributes of items. The framework scores-up recommendations regarding not only the preference of users in items, but also their implicit preference in the attributes of the items. Thus, users might be more likely to click on recommendations offered if these recommendations contain features they know and they are interested in.

Indeed, by using this new knowledge we transform ratings into "semantic values", which better represent the interests of users. Thus, the concept of semantic used in this paper to indicate the expansion in the meaning of ratings. That is, this semantic concept does not lead to inferences or reasonings. A similar idea was used in [6], where authors create a matrix of items-attributes.

Experimentations are done in the domain of movies: we use the large set of ratings in MovieLens and attributes of IMDb database. The results achieved show the good performance of our approach compared to a semantic-less matrix factorization approach.

This article is structured as follows: In Sect. 2 related work is presented. Section 3 explains our approach. In Sect. 4 and Sect. 5 the experimentations and evaluations done are shown. Finally, in Sect. 6 conclusions and possible future work are discussed.

2   Related Work

In general, Recommender Systems (RS) use the feedback of users in items in order to predict their interest in other items. In this state of the art we would like to focus on three aspects of typical RS: (1) the scalability of the system, (2) the capacity of the system to incorporate heterogeneous information, and (3) the domain dependency of the system. Looking for a RS that achieves these goals is not trivial. In [1] presents and explains the paradigms of each recommendation technique. Typical CB techniques can incorporate external heterogeneous information form different resources, but they are domain dependent. CF has demonstrated high accuracy and item domain generality, yet difficult to deal with more heterogeneous data. Other hybrid methods usually combines CB and CF in order to improve recommendations. However, the system increases its domain dependency and complexity, and it becomes more difficult to distribute.

CF techniques based on Matrix Factorization (MF) are specially interesting, since they suit with large amount of data. The Matrix Factorization (MF) technique decomposes a matrix R into two random matrices, P and Q, in such a way that the multiplication of both gives approximately the original one. This concept is used in RS to predict the missing rating values of users using the knowns ones [7]. Indeed, this problem can be resolved by using optimization algorithms. The two most known optimization techniques that may find out accurate predictions are based on alternating minimization and gradient descent [8]. On the one hand, alternating minimization techniques have demonstrated to have a simple algebraical problem resolution. It was popularized by the Alternating Least Square (ALS) method [9,10], and other modifications have been suggested [11,12]. This technique decomposes the problem into two simple optimization problems represented in P and Q. Then, by fixing one matrix, they have to guess the other one. Iterating the fixed matrix in order to guess the other one yields in an approximated result for R. In [12], authors uses the ALS concept to optimize the overall ranking prediction in top-K recommendations. Recently, [13] expose a detailed theoretical discussion about the optimal usability and the accuracy of ALS methods. On the other hand, the gradient descent optimization technique includes learning-parameters that study the ratings patterns to improve the results of the algorithm. It was popularized by [14] and many improvements and variations have been proposed [8,15,16]. In order to minimize the error, this technique iterates among each single entry in R looking for a global minimum. After each iteration, the parameters are updated taking the negative gradient of the function into account. This technique is also known as Singular Value Decomposition (SVD) for recommender systems.

However, these techniques above do not simplify the incorporation of external heterogeneous data. In [8] it is argued that some aspects as the time can be taken into consideration. Yet, still more heterogeneous data can be used to improve the system (e.g. the features of users/items).

In contrast, some authors focus on hybridizations. For instance, [17] suggested a CF and Knowledge-based system to generate multi-type recommendations. A multi-type recommendation suggests not only the goal item, but also some other interesting facts related to the recommended item, e.g. recommending restaurants and the best route to get there. To do that, they use a memory-based CF to compute cosine similarity between experienced cases, and a Case-based Reasoning that adjusts the cases proposed by the CF. Other authors propose Multi-Criteria recommendations [18–22]. Briefly, they consider the ratings from users as a solution for an equation, where the variables are some item attributes. Thus, in order to explain an overall rating in items, they independently analyze explicit ratings given for these attributes and also execute predictions for them. However, these approaches assume the existence of explicit ratings for the attributes of items, but indeed these ratings are hard to get in real-life.

In [23], authors also want to study the interest of users in detail. Their approach uses a three-layer representation, user-interest-item. For a user, an interest is a characteristic that an item must have. For an item, an interest is one of its attributes. Then, they apply a Latent Dirichlet Allocation (LDA) algorithm based on "topic models" from text domains in order to tackle the similar multiple "theme" problem [24]. Hence, authors interpret that the text documents are users, the words are items, and the topics are the latent interests. This extracts hidden interests by establishing a correlation matrix graph about items and interests. This approach shows good performance, although the complexity is not acceptable for large-scale applications.

Other approaches focus on improving the disadvantages of the used recommendation technique. For instance, in [25,26] authors suggest a CF-CB hybrid system to improve recommendations in an item-based collaborative filtering technique. Authors propose a framework to control the similary/diversity factor in a top-K recommended items. The approach is based on clustering techniques.

The most relevant items are hierarchically ordered and forms trees of interest in recommendations, what allows creating a zoom-in technique to see more items of the same tree, which tend to be similar.

The usage of Semantic Technologies may facilitate incorporating heterogeneous data to the system, although it also difficulties its domain independency and its scalability. In [5] author propose a state of art for this topic.

On the one hand, some authors propose creating a profile of the interest of users, like [27,28]. In [27] a hybrid recommender system for TV Programs called Avatar is proposed. It creates structures for both items and users in ontologies and it aims to do inference similarity. Authors use ontologies to implement (1) a content-based technique that computes item similarities, and (2) a collaborative filtering that computes user-profile similarities based on positive and negatives preferences. The system first filters the N most similar user profiles and focus on their positive preferences. After a pre-selection of items that could match to this requirements, they filter out items with negative preferences matches. Finally, they take the top-K items with highest matching values. In the same way, in [28] authors propose a hybrid approach to overcome shortcomings in CB, Knowledge-based and CF. The architecture uses three different agents: Semantic Association Discovery Agent, Data Mining Agent and Random Selection Agent. The former exploits ontologies using knowledge-based techniques to overcome new item problem. The second addresses new user problem. The latter utilizes a random item selection alleviating overspecialization.

On the other hand, other existing approaches focus on better describing items to improve the recommendations. In [29], authors propose a Semantic CB method to improve standard CF techniques. They use item-item similarity based on context pages in Wikipedia to compute "artificial ratings" for an element. These artificial ratings are used instead of classic rating when we have a very sparse user-item matrix.

In contrast, [30] propose to integrate to RS the social network system tags, where users provide keywords. Tags are mapped in concepts within the ontology, bypassing clustering. The approach creates a matrix of items and conceptsn and then, it matches the tags of users to concepts in matrix in order to know adequate items for users. Another approach using keywords and ontologies is [31]. They first characterize items using attributes as keywords. Then, they compute item-similarity regarding their keywords. Besides, they reduce the number of keywords by using WordNet as a concept ontology to establish synonyms or similar meanings among keywords.

In this paper we try to achieve the three goals: a capacity for incorporating heterogeneous information, a high level domain genericity and a scalable system. We suggest a flexible and generic collaborative filtering system that relies on matrix factorization and implicit relations in data. The matrix factorization warranties the scalability and domain genericity of the system. The implicit relations in data allows scoring-up items regarding the implicit interest of users in the attributes of items.

3   Suggested Architecture

In general, recommender systems still can better exploit the feedback of users. This fact may be achieved by improving current recommendation techniques and incorporating external heterogeneous information of users or items [5]. Matrix Factorization techniques have already demonstrated a highly accurate prediction. In addition, it suits with large sets of data and it is domain independent. However, this technique makes difficult the incorporation of heterogeneous data.

In this work we take advantage of the domain independency and the scalability of Matrix Factorization and we try to improve its heterogeneity constraint. We propose to add an external layer, which will be in charge of the external heterogeneous data. In this layer, items, the attributes of items, users and ratings are analyzed together to find out new implicit relations in data. By exploiting this we transfor ratings into "semantic values". Note that the term "semantic" indicates a expansion in the meaning of ratings. Indeed, this new value represents the interests of the users in items and the attributes of the items. Despite this usage, the approach aims to keep a high level of domain independency. As a consequence, in order to achieve presented goals (genericity, scalability and accuracy), we suggest a three-layer recommender architecture: a pre-analysis layer, a semantic layer and a recommender layer. This architecture is shown in the Fig. 1.

Since the number of attributes and the number of values for the attributes might be huge (e.g. all the actors in a movie, or all movie tags), the pre-analysis module implements a feature selection module and a counting module. The former reduces the number of attributes to focus on. The latter speeds up the system while deeply studying the user interests: we count the implicit number of times that a concrete value for an attribute appears among the rated items of users. The semantic module uses the information deduced in the previous layer in order to transform the ratings of users: we expand the meaning of ratings by adding the implicit relation in data. At last, the recommender module uses an existing collaborative filtering technique based on Matrix Factorization technique to generate accurate recommendations and to keep the high scalability and genericity.

3.1   Pre-analysis Module

This layer gathers information from the dataset and the domain description of items (e.g. a database or an ontology), making it abstract and quickly available for next modules. First, we study the relevancy of the attributes in the domain by using Principle Component Analysis (PCA). Then, we analyze the interest of users in these selected attributes and store the deduced information in a fast and low space counting module called Counting Bloom Filter (CBF).

Feature Selection Using PCA. As long as the number of item attributes might be huge, we apply a reduction technique based on PCA to select the

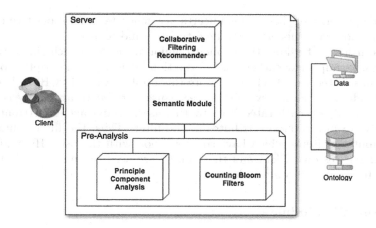

Fig. 1. Global architecture of the recommender system

most representative attributes. Besides, this technique provides weights for such attributes in order to balance their importance in the recommendation. This weight extraction is achieved by deeper studying the results of PCA. One may see how ratings may be explained by the attributes of items.

Counting Module. Ratings reflect the interest of users in items. It is important to understand their item rating-reasons in order to better serve the users. However, an item is composed of several attributes and getting feedback for all of them is complicated. Indeed, users are not willing to rate every single attribute of a movie, e.g. a user may not want to rate every actor in a movie. As a consequence, suggest to implicitly gather this information using the past rated items. For instance, movies with a certain actor might be preferred by users who have rated and liked a movie with this actor. This implicit knowledge should be computed and stored in order to have it quickly available.

On the one hand, databases or semantic technologies as ontologies, describe items environment and they can easily return their unique properties. This fact gives free access to navigate through items features. On the other hand, the implicit information should be stored to have this information quickly accessible. We use using Counting Bloom Filter to this fact.

A Bloom Filter (BF) is a bit structure that allows to represent a set of elements in a lower space size [32]. It uses hash functions in order to efficiently distribute elements among the structure. This filter allows doing fast membership queries, and hence, one may check whether an element is in the structure or not (presence or absence). However, it can not say how many times an element appears in the filter. Counter Bloom Filter (CBF) addresses this constraint by adding counter bits to the filter [33]. These filters have a limitation while doing membership queries: filters assure the absence of an element, but they do not assure the presence of them. This uncertainty generates a false positive ratio to deal with. However, this error can be estimated and reduced.

Thus, instead of asking the explicit opinion of users in each single value of attributes, we implicitly gather this information by using the description of items and the past rated items. Then we store this implicit value in CBFs. The steps of this module are as follows: (1) for each user we create an empty counting bloom filter, (2) for each rated item by this user, we extract its attributes and (3) finally we insert these attributes in the filter. Thus, the filter contains all the attributes of items which have some relation with the user. Highlight that each user has his own CBF, and these filters are used by the semantic module in order to improve recommendations.

3.2 Semantic Module

This module aims to expand the meaning of a rating by incorporating the implicit interest of users in the attributes of items. As said above, an item is composed of several attributes and getting feedbacks for all of them is complicated. The CBF of a user contains the implicit interest of the user in the attributes of an item. We aim to exploit this information in order to add a new sense to users feedback. This expands the meaning of ratings, what we dubbed "semantic values".

The semantic module transforms the initial rating given by users into a new "semantic rating". Indeed, this new value takes into account not only the user preference in the item but also the preference in the attributes of the item. For instance, an item rated as 4 out of 5 may transform its rating value into 4.5. This fact reflects that this item has several attributes in common with the rest of items rated by the user. As a consequence, this boosts the recommendation of items which contain similar attributes to the ones the user liked in the past. Hence, recommended items are more suitable and acceptable by users because they may recognize relevant features for them.

The transformation of the ratings follows the equation presented in (1). Equally, we call it "semantic equation" because it aims to expand the sense of a rating.

$$sv_{u,i} = r_{u,i} + E[r_{u,*}] * \frac{\left| \sum_{j=1}^{F} C_j * W_j \right|}{N_u} \tag{1}$$

Here, $r_{u,i}$ is the real rating for item "i" given by user "u". N_u is the total number of items rated by user "u". $E[r_{u,*}]$ is the average of the ratings given by user "u". F is the number of selected attributes. W_j are the weights for these attributes computed by PCA. C_j are the number of times that the value of an attribute has appeared for a user, easily got using the computed CBF. Besides, since parameters are pre-calculated, the number of attributes does not have a relevant impact on the execution time of the module. In addition, the process of this equation is easy to parallelize.

Moreover, we use this equation in two different levels of the recommendation. On the one hand, we apply it to all the ratings available in the original training dataset, which is the input approach. On the other hand, we apply the semantic

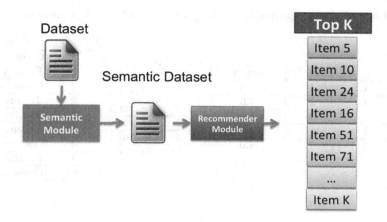

Fig. 2. Semantic dataset: input approach

equation to the output of the recommendation. These approaches are explained in the next Subsects. 3.3 and 3.4.

3.3 Semantic Dataset (Input Approach)

This approach implements the semantic module at the input of the recommender module. Briefly, it transforms feedback in the training dataset into a semantic feedback, according to the semantic equation (1). That is, for each rating a new "semantic rating" is computed. Hence, a "semantic dataset" is built from the original one. The Fig. 2 shows this approach. The semantic module takes a training dataset, which contains the "original dataset", and generates a new "semantic dataset", which contains the new "semantic ratings". This latter is used to train the recommender module and create a prediction model to exploit. As the incoming dataset has changed, the recommendation module can return different items.

Remember that collaborative filtering analyzes the ratings of users in order to find out patterns to group similar users. In this approach the recommendation technique still looks for similarities among users, by involving not only items but also attributes. In fact, by increasing the ratings of items in which users are interested (according to their interest in the attributes of items), one helps the recommendation technique to focus on such accuracy and predictions. As a result, the latent space model created by Matrix Factorization learn the importance of these items.

Example. Imagine a reduced dataset as shown in the set of Table 1. It contains information about ratings of users in items and the attributes of items, in this case in the domain of movies (genres and actors). This approach takes and modifies every rating in the dataset according to the implicit interest of users in the attributes of items.

Table 1. Example. (a)Ratings table, (b)Actors table and (c)Genres tables

User "u"	Movie "i"	Rating "$r_{u,i}$"
1	1	4.0
1	2	3.0
1	3	1.0
1	4	2.0
2	9	4.0

Movie "i"	Actor
1	Actor 1
1	Actor 3
2	Actor 1
3	Actor 2
4	Actor 1
10	Actor 3

Movie "i"	Genre
1	Comedy
1	Fantasy
2	Comedy
2	Drama
3	Thriller
3	Drama
4	Comedy
4	Fantasy
10	Comedy

Focus on the rating of the item 1 given by the user 1 ($r_{u,i} = r_{1,1} = 4$). Our goal is to obtain a new "semantic rating" for this value. We first calculate the average of ratings for this user, who has rated $N_u = 4$ movies:

$$E[r_{1,*}] = \frac{4.0 + 3.0 + 1.0 + 2.0}{4} = 2.50 \tag{2}$$

Secondly, we get the weight for attributes computed by PCA (e.g. $W_1 = 0.4$ and $W_1 = 0.6$ for genres and actors respectively). The third step is to get the implicit occurrences stored in CBF:

- The user 1 has rated the items 1, 2, 3 and 4, and these items have actors and genres.
- Focus on the item 1 and its genres: comedy and fantasy. Already rated movies 2 and 4 are comedies, besides the movie 4 is also a fantasy movie. Hence, the occurrences count $C_1 = 3$.
- Focus on the item 1 and its actors: actor 1 and actor 3. The actor 1 also appears on movies 2 and 4. Thus, the occurrences count in this attribute $C_2 = 2$, since the actor 3 does not appear on any other movie.

Putting everything into the equation, we obtain the new "semantic rating":

$$sv_{1,1} = 4.0 + 2.50 * \frac{|3 * 0.4 + 2 * 0.6|}{4} = 5.5 \tag{3}$$

3.4   Semantic Top-K (Output Approach)

Recommendations given by this collaborative filtering are pertinent due to its collaborative nature: it analyzes the ratings of users in order to find out patterns to group similar users. However, users may prefer some features in movies rather than others, e.g. a movie has a high rating because the user like the actor independently of the genre of the movie. This approach implements the

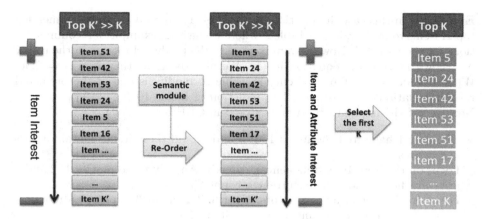

Fig. 3. Semantic top-K: output approach

semantic module at the output of the recommender system. It modifies the recommendations done by an already built collaborative filtering recommender in order to insert the interest of users in the attribute of items. This adapts the recommendations to users based on his implicit feedback in the features of items.

For the user to whom recommendations are required, it takes the top-K' $(K' \gg K)$ best predictions from the recommender system and transforms the rating predictions of these recommendations. This prediction modification aims to better adapt the recommendations to the users. Indeed, it takes into account the singular preferences of the user in the attributes of the items.

Each item in the top-K' contains a predicted rating which reflects the interest in the item. In fact, this top-K' is usually ordered by this predicted value, hence items in the top are likely more interesting for users. We aim to change this prediction into a "semantic prediction". For this purpose, two sets are required: (1) items in a top-K', and (2) the preferences of the user, i.e. the set of rated items by the user. The former is the recommended item which has attributes and a predicted interest value. The latter implicitly contains attributes in which the user is interested in. In this conditions, we apply the semantic module to change the prediction value. Doing this process among the whole top-K' results in a new resorted top-K', which contains the same items in different positions. Thus, we score up items with similar attributes to the ones the user is interested in. The fact of taking $K' \gg K$ helps the system to put in the top-K new relevant items which initially were out of it. Finally, the system returns the smaller top-K new best items of the re-ordered top-K'. The Fig. 3 represents this process.

This approach is much faster than the semantic dataset because it requires to transform many less ratings. In addition, since collaborative filtering uses to return a certain grade of diversity in their predictions [1], we adjust the top-K items according to the interest in items and attributes.

Example. In this case, in addition to the Table 1, we need also a recommended top-K items to modify, like in Table 2, where we have an example of recommendations for the user "1". Now, we aim to modify the predicted ratings in the top-K by using the semantic equation in (1). We already know that $E[r_{1,*}] = 2.50$, $W_1 = 0.4$ and $W_2 = 0.6$. The current value to modify is one of the predicted recommendations, for instance the prediction of movie 10 ($r_{u,i} = r_{1,10} = 4.5$). Now we get the implicit occurrences stored in CBF:

- The user 1 has rated the items 1, 2, 3 and 4, and these items contain actors and genres.
- Focus on the item 10 and its genres: comedy. Already rated movies 1, 2 and 4 are comedies. Hence, the occurrences count $C_1 = 3$.
- Focus on the item 1 and its actors: actor 3. The actor 3 appears on movie 3. Thus, the occurrences count in this attribute $C_2 = 1$.

Putting everything into the equation, we obtain the new "semantic rating":

$$sv_{1,10} = 4.5 + 2.50 * \frac{|3 * 0.4 + 1 * 0.6|}{4} = 5.625 \tag{4}$$

Applied to the whole top-K', this process provokes a new order in the top-K. This new recommendations are more personalized to the user according to the interest in the attributes of items.

Table 2. Example. Top-3 recommendations for the user "1"

Top-3	Movie 21	Movie 10	Movie 64
Predicted Rating	5	4.5	4

4   Experimentation

Dataset

We suggest using the ratings in MovieLens dataset[1] and domain attributes from IMDb[2] database. This merged dataset is provided by GroupLens [34]. It is composed of 2113 users and 855598 ratings over 10197 movies. It also offers six attributes: genre, directors, actors, countries, locations and tags. The total number of distinct values for these attributes is 112881: 20 movie genres, 95321 actors, 72 countries, 4266 locations and 13222 tags. Under the authors knowledge, there is not any public and available ontology that perfectly fits in this dataset. Thus, for experimentation purposes, the ontology relations are modeled within a sql database, as done in [6].

---

[1] http://grouplens.org/datasets/movielens.
[2] http://www.imdb.com/.

Table 3. Experimentation: Weights % for variables in dimensions. Approximate values.

Variables	Actor	Country	Director	Genre	Location	Total
Dimension D_1	19.537	12.719	19.896	0.000	5.064	57,25
Dimension D_2	4.785	8.3732	5.303	6.459	17.823	42.75
Total (%)	24	21	25	6	23	100

Principle Component Analysis

Due to the high number of ratings in the MovieLens dataset, and in order to apply the feature selection, we extract the 100 users who have rated the highest number of movies. Thus, we obtain 169155 ratings, which represent almost the 19.77 % of the total ratings in MovieLens dataset. The PCA method analyzes the relevancy of items attributes over this data and returns the most representative features. In addition, it returns relevancy ceiled-weights for these attributes. As is shown in Table 3, this module takes out the attribute "tags" since it seems to be, for the PCA, the less relevant over the presented ones.

Counting Bloom Filter

CBF are built in off-line in order to speed up the semantic equation. The dataset contains 2113 users and 112881 different values for the attributes. Regarding the CBF structure, we accept a very low false-positive ratio of 0.01 %. In addition, we consider that each value for each user will not appear more than 64 times. That is, we set 6 bits for counting tasks. As a result, the size of one filter corresponding to one single user is around 1.3 Mb. Hence, for the 2113 users the total size of all filters is around 2.7 Gb.

Recommender

We use the SVD algorithm in Apache Mahout[3] to build the recommender core. This algorithm will iterate a maximum of 30 times to find out the best 30 latent-features that explain the ratings. However, the semantic module uses this recommender as a black box. The experimentations are done in both explained configurations: semantic dataset and semantic top-K approaches.

5   Evaluation and Results

Our approach uses the features of items and the past preferences of users in order to get a new hidden implicit information about the interest of users in these features. This fact does not directly affect the recommendation. In fact, item

---

[3] https://mahout.apache.org.

similarity measures or items comparisons are not considered, and hence there is not any content-based techique used. Indeed, we exploit these implicit analysis to enhance collaborative filtering recommendations. Due to this assumption, we do not consider our approach a hybrid method: the core of recommendations remains a pure collaborative filtering technique. Because of that, we would like to compare the behavior of our "semantic" recommender system approaches against a non-semantic system.

We aim to study the behavior of the system regarding the ratings in training data. The more training data, the better one can profile a user, and thus, the better one study the implicit interest of users in the attributes of items. In fact, this dependency on the training data corresponds to a study of different sparsity levels. Therefore, for the evaluation of the systems we use the full MovieLens dataset containing 855598 ratings over 10197 movies. To represent the different sparsity levels, we randomly split the dataset into 90 %, 80 %, 70 %, 60 % and 50 % training sets. The remaining percentage in each level is the test set[4]. As a consequence, we can train systems and compare the predictions in the model with the real-observed values in the test set.

In order to demonstrate the properties of the approaches, we use three different evaluations: a prediction accuracy based one, a ranking accuracy one and an item similarity evaluation.

Finally, note that the graphs show the results of three approaches: SVD, semantic dataset and semantic top-K. The former is the semantic-less recommender system. The second implements the semantic at the input of the system. The last uses the semantic at the output of the recommender module.

Outline

This outline aims to give a deployed example of what the recommender systems return. It visually compares top-K returned items from the different approaches. The interest of this outline is to compare the items that different recommenders may show to the same user.

We focus on the user 6757, who is the user with more ratings and hence the best profiled user. (1) In the training set, we subtract 60 out of 119 ratings with the maximum rating score (movies rated with a 5). (2) Then, we train the three different systems in this context: the SVD approach creates a model using this training set, the Semantic Dataset approach first apply the semantic equation to the training set and then creates a model, and the Semantic Top-K approach modifies the recommendations done by the simple SVD approach. (3) Finally, we ask the systems for a top-60 items for user 6757, expecting to find those ratings deleted from the training set. Table 4 shows the top-10 items (over these 60 movies).

---

[4] Denote that, since the convergence of the collaborative filtering has been already proved and the semantic approaches do not modify this convergence capability, we do not need a cross-validation set.

Table 4. Experimentation: Top-10 recommendation for user 6757. Items ID and predicted values

SVD		Sem. Dataset		Sem.top-K	
ID	Value	ID	Value	ID	Value
6669	4.34	6669	4.21	6669	4.40
26350	4.20	858	4.14	912	4.31
858	4.189	912	4.13	858	4.239
912	4.186	26350	4.09	8492	4.237
8492	4.16	7749	4.08	26350	4.23
7762	4.128	1221	4.07	3462	4.226
3077	4.1241	3462	4.05	2624	4.224
7749	4.1240	7762	4.03	4806	4.219
4806	4.12	1213	4.027	1221	4.218
2624	4.11	8492	4.026	7256	4.210

The semantic-less recommendations returns 2 items (858 and 912) which belong to the extracted items. However, the semantic approaches improve this fact: the semantic dataset returns 4 items (858, 912, 1213 and 1221) and the semantic top-K returns 3 items (858, 912 and 1221). This fact is due to the accuracy of the SVD and the extra-knowledge added by the implicit interest in features of items. In addition, we notice the appearance of different items in the semantic approaches (such as item 3462). Specially, we highlight new order in items of the semantic top-K (items 912 or 2624). In fact, we have scored up items which contain interesting attributes for the user, and thus, less interesting items regarding attributes get down in the list. These results show our assumptions: by adding the implicit interest of users in items, recommendations are more suitable and acceptable to users, i.e. more items out of the extracted high scored items set are predicted.

Root Mean Square Error (RMSE)

The RMSE measure evaluates the system in terms of accuracy of the ratings prediction. It represents the standard deviation in the error of the prediction. This error is the difference between predicted values and real-observed values in the test set. Thus, the lower is this error, the better is this metric.

Since our frameworks modify the ratings, they do not overcome the accuracy of the SVD. The reason is that the semantic module scores up items due to the presence of attributes, yet it does not penalize the absence of them. Thus, the semantic rating is always higher than the explicit ratings. These results are shown in Fig. 4. A further study is being doing to improve this fact.

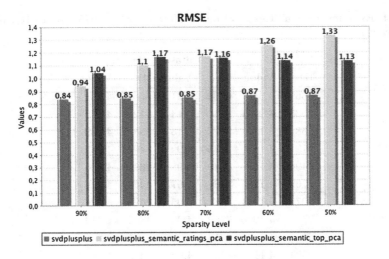

Fig. 4. RMSE metric comparisons.

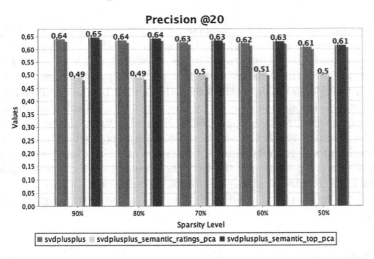

Fig. 5. Precision metric comparisons regarding a top-20 items.

Precision, Recall and F-Measure

Precision and Recall techniques measure the relevancy of items in a previously selected top-K. This relevancy is a binary value associated to the item: an item is relevant or not regardless its predicted rating value. Precision represents the percentage of relevant items (items that should be recommended first) over the recommended top-K items. Recall represents the percentage of relevant items over the whole set of items. Figures in 5 and 6 show the results in precision and recall measures.

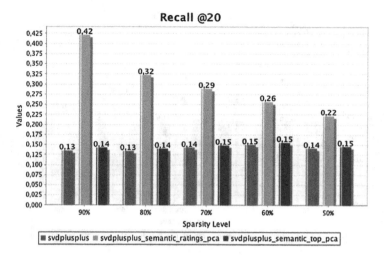

Fig. 6. Recall metric comparisons regarding a top-20 items.

On the one hand, due to the prediction accuracy of the SVD, the non-semantic system puts easily relevant items in the ranking, and thus precision is high. However, the semantic top-K approach slightly overcomes this precision, since it scores-up items and thus relevant items are likely to appear. On the contrary, in the semantic dataset approach, the ratings modification affects to this accuracy and thus precision is fewer. On the other hand, since we score up items which contain interest attributes for the users, our semantic approaches identify more relevant items among the whole dataset, and hence recall metric are higher, specially in the case of "semantic dataset".

The F-Measure and the F2-Measure are figure of merits for Precision and Recall. The former equally balance the importance of precision and recall. The latter gives the double importance to precision than to recall. Figures 7 and 8 show these metrics. One may observe that by adding a semantic layer improves top-K recommendations, enhancing the overall performance of the system as well. Summing up, scoring up items with common attributes indeed increases the probability of taking relevant items.

Intra-List-Similarity (ILS) and Intra-List-Diversity (ILD)

Recommending always too similar items may bore users, and too different items might generate confusion. The ILS (Intra-List-Similarity) metric, also called ILD (Intra-List-Diversity), balances items similarity/diversity among a recommended top-K. In a scale 0–1, the closer is the value to "1", the more similar items in the top-K are between them. On the contrary, the closer to "0", the more diversity exists among recommendations. Typically, collaborative filtering technique tends to show some diversity among its recommendations. Adding a semantic layer either at the input or at the output of the system, one increases the similarity

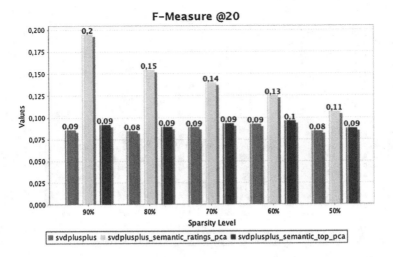

Fig. 7. F-Measure metric comparisons regarding a top-20 items.

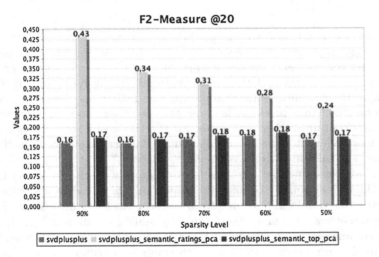

Fig. 8. F2-Measure metric comparisons regarding a top-20 items.

among recommendations. This added similarity is indeed based on the interest of users in the attributes of items. These facts are shown in Figs. 9 and 10, where we represent the similarity/diversity measure regarding the genre attribute and the actor attribute of items.

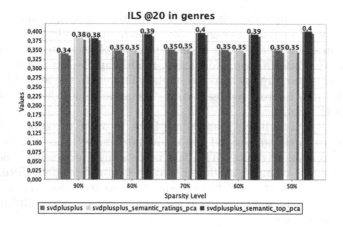

Fig. 9. ILS metric comparisons regarding the attribute "genre".

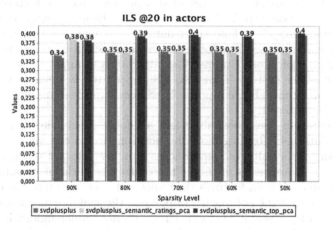

Fig. 10. ILS metric comparisons regarding the attribute "actor".

6   Conclusion

Recommender systems select, among a huge amount of data, the information in which users might be more interested. In order to do that, these systems exploit the known interest of users in items, which is in an explicit or implicit feedback. In this paper, we highlighted a lack of feedback regarding the attributes of items, which may be really useful for improving recommendations. However this information is hard to retrieve: users are not willing to rate all aspects of items (e.g. all the actors in a movie).

We proposed an approach which relies on collaborative filtering techniques and implicit relations in data. On the one hand, using CF techniques based on MF, one can generate very accurate recommendation in a parallelizable algorithm. Besides, this fact alleviates the analysis of large datasets. On the other hand, the description of items allows making more relations among data. Thus, one can easily extract the implicit interest of users in the attributes of items. Using this information, we suggest modifying the explicit ratings given by users in order to represent also the implicit interest in the attribute of items. We called this new interest representation "semantic value", because it expands the meaning of ratings.

The process is as follows: first we use PCA in order to reduce the number of attributes to focus on. Second, we count how many times a user has liked an item with certain attributes. This fact needs a high processing time which is reduced by using a counting module based on Counting Bloom Filters (CBF). Third, we use this new stored data to modify the ratings of user in items. Finally, new updated recommendations are done using a collaborative filtering matrix factorization technique

The presented architecture is divided in independent layers and allows a flexible usage. Two approaches are presented regarding this architecture: Semantic Dataset and Semantic Top-K. The former acts in the input of the recommender system by analyzing the whole train dataset and modifies the input ratings. Enhancing the relevancy of attributes in the feedback of users, we help the system to focus on such kind of items. The latter aims to apply the semantic layer in the output of the system. Typically, RS provide top-K items ordered by predicted user's preference. In this approach, we score-up the items whose attributes may be of users interest.

Experimentation uses the public and available MovieLens dataset merged with IMDb database. Results show the performance of the approach over different measures. Specially, our approaches enhance the fact of taking relevant items for users. Thus, users might be more likely to click on recommendations because they may contain features they know and they are interested in.

Finally, note that the approaches implement the semantic technologies taking into account process scalability and a high domain independent level. Future work focuses on the penalization in the appearance of non-preferred attributes and on the agility in the counting structure.

References

1. Ricci, F., Rokach, L., Shapira, B., Kantor, P.B. (eds.): Recommender Systems Handbook. Springer, US (2011)
2. Breese, J.S., Heckerman, D., Kadie, C.: Empirical analysis of predictive algorithms for collaborative filtering. In: Proceedings of the Fourteenth Conference on Uncertainty in artificial intelligence, Morgan Kaufmann Publishers Inc., pp. 43–52 (1998)
3. Su, X., Khoshgoftaar, T.M.: A survey of collaborative filtering techniques. Adv. Artificial Intell. **2009**, 4 (2009)
4. Koren, Y.: The bellkor solution to the netflix grand prize. Netflix prize documentation (2009)
5. Peis, E., del Castillo, J.M., Delgado-López, J.: Semantic recommender systems. analysis of the state of the topic. Hipertext. net **6**, 1–5 (2008)

6. Mobasher, B., Jin, X., Zhou, Y.: Semantically enhanced collaborative filtering on the web. In: Berendt, B., Hotho, A., Mladenič, D., van Someren, M., Spiliopoulou, M., Stumme, G. (eds.) EWMF 2003. LNCS (LNAI), vol. 3209, pp. 57–76. Springer, Heidelberg (2004)

7. Koren, Y., Bell, R., Volinsky, C.: Matrix factorization techniques for recommender systems. Computer **42**(8), 30–37 (2009)

8. Koren, Y., Bell, R.: Advances in collaborative filtering. In: Ricci, F., Rokach, L., Shapira, B., Kantor, P.B. (eds.) Recommender Systems Handbook, pp. 145–186. Springer, US (2011)

9. Schafer, J.B., Frankowski, D., Herlocker, J., Sen, S.: Collaborative filtering recommender systems. In: Brusilovsky, P., Kobsa, A., Nejdl, W. (eds.) Adaptive Web 2007. LNCS, vol. 4321, pp. 291–324. Springer, Heidelberg (2007)

10. Zhou, Y., Wilkinson, D., Schreiber, R., Pan, R.: Large-scale parallel collaborative filtering for the netflix prize. In: Fleischer, R., Xu, J. (eds.) AAIM 2008. LNCS, vol. 5034, pp. 337–348. Springer, Heidelberg (2008)

11. Pilászy, I., Zibriczky, D., Tikk, D.: Fast als-based matrix factorization for explicit and implicit feedback datasets. In: Proceedings of the fourth ACM conference on Recommender systems, pp. 71–78. ACM (2010)

12. Takács, G., Tikk, D.: Alternating least squares for personalized ranking. In: Proceedings of the sixth ACM conference on Recommender systems, pp. 83–90. ACM (2012)

13. Jain, P., Netrapalli, P., Sanghavi, S.: Low-rank matrix completion using alternating minimization. In: Proceedings of the 45th annual ACM Symposium on Symposium on Theory of Computing, 665–674. ACM (2013)

14. Funk, S.: Netflix update: Try this at home (3rd place) (2006)

15. Takács, G., Pilászy, I., Németh, B., Tikk, D.: Major components of the gravity recommendation system. ACM SIGKDD Explor. Newsl. **9**(2), 80–83 (2007)

16. Koren, Y.: Factorization meets the neighborhood: a multifaceted collaborative filtering model. In: Proceedings of the 14th ACM SIGKDD International Conference on Knowledge discovery and data mining, pp. 426–434. ACM (2008)

17. Zhuo, G., Sun, J., Yu, X.: A framework for multi-type recommendations. In: Eighth International Conference on Fuzzy Systems and Knowledge Discovery (FSKD), vol. 3, pp. 1884–1887. IEEE (2011)

18. Adomavicius, G., Kwon, Y.: New recommendation techniques for multicriteria rating systems. IEEE Intell. Syst. **22**(3), 48–55 (2007)

19. Li, Q., Wang, C., Geng, G.: Improving personalized services in mobile commerce by a novel multicriteria rating approach. In: Proceedings of the 17th International Conference on World Wide Web. WWW 2008, pp. 1235–1236. ACM, New York (2008)

20. Lakiotaki, K., Tsafarakis, S., Matsatsinis, N.: UTARec: a recommender system based on multiple criteria analysis. In: Proceedings of the 2008 ACM Conference on Recommender Systems, pp. 219–226. ACM (2008)

21. Mikeli, A., Apostolou, D., Despotis, D.: A multi-criteria recommendation method for interval scaled ratings. In: IEEE/WIC/ACM International Joint Conferences on Web Intelligence (WI) and Intelligent Agent Technologies (IAT), vol. 3, pp. 9–12 (2013)

22. Mikeli, A., Sotiros, D., Apostolou, D., Despotis, D.: A multi-criteria recommender system incorporating intensity of preferences. In: Fourth International Conference on Information, Intelligence, Systems and Applications (IISA), pp. 1–6 (2013)

23. Liu, Q., Chen, E., Xiong, H., Ding, C.H., Chen, J.: Enhancing collaborative filtering by user interest expansion via personalized ranking. IEEE Trans. Syst. Man Cybern. Part B Cybern. **42**(1), 218–233 (2012)

24. Blei, D.M., Ng, A.Y., Jordan, M.I.: Latent dirichlet allocation. the. Journal of machine Learning research **3**, 993–1022 (2003)

25. Boim, R., Milo, T., Novgorodov, S.: Direc: Diversified recommendations for semantic-less collaborative filtering. In: IEEE 27th International Conference on Data Engineering (ICDE), pp. 1312–1315. IEEE (2011)

26. Boim, R., Milo, T., Novgorodov, S.: Diversification and refinement in collaborative filtering recommender. In: Proceedings of the 20th ACM International Conference on Information and Knowledge Management, pp. 739–744. ACM (2011)

27. Fernández, Y.B., Arias, J.J.P., Nores, M.L., Solla, A.G., Cabrer, M.R.: Avatar: an improved solution for personalized tv based on semantic inference. IEEE Trans. Consum. Electron. **52**(1), 223–231 (2006)

28. Pan, P.Y., Wang, C.H., Horng, G.J., Cheng, S.T.: The development of an ontology-based adaptive personalized recommender system. In: International Conference on Electronics and Information Engineering (ICEIE), vol. 1, pp. V1–76. IEEE (2010)

29. Katz, G., Ofek, N., Shapira, B., Rokach, L., Shani, G.: Using wikipedia to boost collaborative filtering techniques. In: Proceedings of the fifth ACM conference on Recommender systems. ACM, pp. 285–288 (2011)

30. Mabroukeh, N.R., Ezeife, C.I.: Ontology-based web recommendation from tags. In: IEEE 27th International Conference on Data Engineering Workshops (ICDEW), 2011. IEEE, pp. 206–211 (2011)

31. Werner, D., Cruz, C., Nicolle, C.: Ontology-based recommender system of economic articles (2013). arXiv preprint arxiv:1301.4781

32. Bloom, B.H.: Space/time trade-offs in hash coding with allowable errors. Commun. ACM **13**(7), 422–426 (1970)

33. Broder, A., Mitzenmacher, M.: Network applications of bloom filters: a survey. Internet Math. **1**(4), 485–509 (2004)

34. Cantador, I., Brusilovsky, P., Kuflik, T.: 2nd workshop on information heterogeneity and fusion in recommender systems (hetrec 2011). In: Proceedings of the 5th ACM conference on Recommender systems. RecSys 2011,. ACM, New York (2011)

Semantic Web-Based Social Media Analysis

Liviu-Adrian Cotfas[1,2(✉)], Camelia Delcea[2], Antonin Segault[1],
and Ioan Roxin[1]

[1] Franche-Comté University, Montbéliard, France
{liviu-adrian.cotfas,ioan.roxin}@univ-fcomte.fr,
antonin.segault@edu.univ-fcomte.fr
[2] Bucharest University of Economic Studies, Bucharest, Romania
camelia.delcea@csie.ase.ro

Abstract. With the on growing usage of microblogging services, such as Twitter, millions of users share opinions daily on virtually everything. Making sense of this huge amount of data using sentiment and emotion analysis, can provide invaluable benefits to organizations trying to better understand what the public thinks about their services and products. While the vast majority of now-a-days researches are solely focusing on improving the algorithms used for sentiment and emotion evaluation, the present one underlines the benefits of using a semantic based approach for modeling the analysis' results, the emotions and the social media specific concepts. By storing the results as structured data, the possibilities offered by semantic web technologies, such as inference and accessing the vast knowledge in Linked Open Data, can be fully exploited. The paper also presents a novel semantic social media analysis platform, which is able to properly emphasize the users' complex feeling such as happiness, affection, surprise, anger or sadness.

Keywords: Ontology · Emotion analysis · Sentiment analysis · Semantic web · Twitter · Social media analysis · Opinion mining

1 Introduction

The last few years have witnessed an amazingly fast-paced growth in the usage of social media networks. Thus, the most commonly used micro-blogging service, Twitter[1], which allows users to broadcast 140 character status messages, also known as tweets, has over 240 million monthly active users, who post more than 500 million tweets every day, as reported in April 2014. Many of these messages contain sentiment and emotion indications regarding almost any topic, therefore turning Twitter into a rich data source for analyzing the public's opinion. Moreover, various studies have already shown that users frequently express opinions regarding products and services in their tweets [1]. These opinions were found to highly influence other consumers' buying decisions as shown in [2]. Therefore, correctly extracting the users' point of view could easily provide invaluable information to companies willing to better understand their customers' needs. Compared to traditional marketing studies, which

---

[1] http://www.twitter.com.

© Springer-Verlag Berlin Heidelberg 2016
N.T. Nguyen and R. Kowalczyk (Eds.): TCCI XXII, LNCS 9655, pp. 147–166, 2016.
DOI: 10.1007/978-3-662-49619-0_8

can take time and involve high costs, social media emotion and sentiment analysis offers the promise of obtaining almost real-time opinions from huge numbers of actual or potential customers.

Sentiment and emotion analysis are growing areas of Natural Language Processing, commonly used to get insights from customer reviews, blogs and more recently from social media messages. They require a multidisciplinary approach, combining elements from fields such as linguistics, psychology and artificial intelligence. Among the tasks to which they have already been applied, we can mention analyzing customers' opinions [3–5], analyzing public's opinion during crisis [6], predicting political elections outcome [7], evaluating the performance of healthcare companies [8] and even stock market evolution prediction [9].

Sentiment analysis is used to determine whether a text expresses a positive, negative or neutral perception [10, 11], also known as polarity. Besides simply determining the perception, some papers also investigate how the strength of the perception should be evaluated [12], thus providing a more in-depth understanding of the user's actual feelings.

While knowing the perception of the user is definitely important, analyzing the categories of emotions contained in Twitter messages using emotion analysis can provide even more information, by putting the focus on the actual feelings, such as joy, surprise, sadness or anger.

Aspect based sentiment and emotion analysis are able to complete the picture by associating the determined perceptions and emotions with particular properties of the analyzed entities, thus taking into account the fact that users frequently express different and sometimes event contradictory feelings regarding the various features and characteristics of a product or service, also called facets [13]. Detailed surveys of exiting sentiment and emotion analysis approaches are presented in [14–16].

While various social media sentiment and emotion analysis approaches have been proposed and evaluated in the scientific literature, only a few papers partially address an equally important aspect, represented by the manner in which the extracted tweets, together with their associated data, the analyzed entities and their facets, as well as the results of the analysis could be stored in a standardized, easily interchangeable and extensible way. Semantic web technologies cannot only meet these requirements, but also, by employing techniques such as interlinking and reusing of classes and properties from well-known ontologies [17], bridges towards the huge amount of knowledge available in Linked Open Data can be created. Therefore, innovative social media analysis platforms can be developed, capable of providing increasingly deeper insights into the customer's real opinions.

In this paper, an end-to-end semantic approach – TweetOntoSense - is proposed, that uses ontologies to model the various emotions expressed in social media messages, the analyzed entities and their facets, the results of the analysis, as well as the various Twitter related concepts. To the best of our knowledge, there are no current scientific publications or commercial systems proposing a fully semantic social media analysis approach. The other contributions of the paper are the TweetOntoSense ontology and the Twitter ontology. By storing the extracted information as triples, advanced analysis can be performed using the technologies associated with the semantic web.

The paper is organized as follows. In the Sect. 2, a survey of existing ontology based approaches found in the scientific literature is provided. The Sect. 3 presents the Emotions, Twitter and TweetOntoSense ontologies, which form the bases of the proposed approach. The Sect. 4 of the paper includes the steps needed to perform sentiment and emotion analysis. The Sect. 5 shows how the extracted information can be further exploited using semantic web inference and SPARQL (SPARQL Protocol and RDF Query Language) queries, to create the bases for developing an advanced social media analysis platform. The Sect. 6 summarizes the paper and introduces possible future research directions.

2 Ontology Based Approaches

According to [18], ontologies are defined as a "formal, specification of a shared conceptualization". They formally represent knowledge as a hierarchy of concepts, using a shared vocabulary to denote the types, properties and interrelationships of those concepts. Currently, ontologies have become the means of choice for representing knowledge, by both providing a common understanding for concepts and being machine processable.

Existing ontology based social media sentiment and emotion analysis approaches can be classified in respect to the usage of ontologies in:

- approaches modelling only the sentiments and emotions;
- approaches modelling only the analyzed entities and their facets.

2.1 Approaches Modelling Only the Sentiments and Emotions

An approach for modelling the sentiments and emotions found in Twitter messages is proposed in [19].

The ontology includes seven basic emotions, composed from the six Ekman emotions and the additional "love" emotion. The emotions are structured in the positive, negative and unexpected categories, as shown in Fig. 1, corresponding to the possible sentiment polarities.

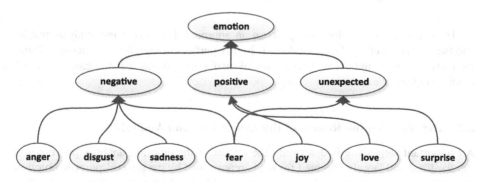

Fig. 1. Ontology for emotion representation used in [19]

A potential downside of the approach is represented by the limited set of emotions, which might not be able to capture all the shades of the opinions expressed in the social media messages. As shown in the Sect. 3 of the paper, our approach relies on a more complex ontology, that structures emotions in a multi-level hierarchy, in which with every level, emotions become more and more fine-grained.

2.2   Approaches Modelling the Analyzed Entities and Their Facets

Such approaches take into consideration the fact that users express opinions about the various characteristics of the analyzed entities and not only about the product or service as a whole. Modelling the analyzed entities and their facets using an ontology is investigated in [3], where the authors show how this approach could be applied for evaluating the public's sentiments regarding the different characteristics of several popular smartphones. An extract from the proposed ontology is presented in Fig. 2.

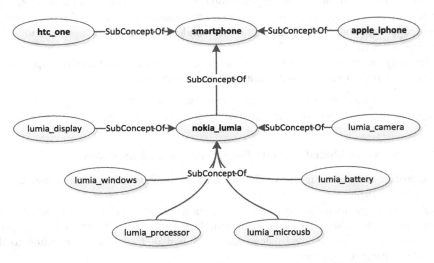

Fig. 2. Ontology for object – facet representation used in [3]

However, the paper does not propose an approach for storing the analysis results and the described entity-facet ontology is highly particular, therefore not being suitable for more general semantic web-based social media analysis, as it is the case for social media analysis platforms.

2.3   Representing the Results of the Aspect Emotion Analysis

An important step towards describing the results of the emotion analysis process in a standardized and largely accepted format is represented by the general-purpose emotion annotation and representation language Emotion Markup Language – EmotionML [20].

It is a W3C recommendation for representing emotion related states in data processing systems and provides twelve vocabularies for appraisals, categories and dimensions, further described in [21].

An ontology-based approach for representing sentiment analysis results is represented by Marl [22], a vocabulary designed to annotate and describe subjective opinions expressed on the web. The Onyx ontology is a recent development towards representing emotion analysis results, of the approach proposed in Marl. It aims to provide a simple means to describe emotion analysis processes and results using semantic web technologies [15]. It is organized around the *onyx:EmotionAnalysis*, *onyx:EmotionSet* and *onyx:Emotion* classes and reuses several properties and classes, such as *prov:Activity* and *prov:Entity*, from the W3C Provenance Ontology [23]. Neither Marls, nor Onyx are able to provide a complete description for both sentiment and emotion analysis results. Given the fact that they were not specifically designed for performing aspect-based social media messages analysis, they cannot represent important information, such as the analyzed entity, it's facets, the Twitter account, its followers and much more additional data, that is highly relevant in the context of social media analysis.

3   Twitter Sentiment and Emotion Analysis Ontologies

The concepts needed in order to perform aspect-based sentiment and emotion analysis on Twitter messages can be grouped in three main categories:

- concepts that express human emotions;
- concepts that describe Twitter specific knowledge;
- concepts that provide a connection between the Twitter message, the expressed sentiments, the expressed emotions and the analyzed entity and its facets.

As shown in the previous section, while various existing researches focus on the different components required for building a fully semantic approach for aspect emotion and sentiment social media analysis, currently there are no end-to-end semantic solutions.

While for the first category of concepts, the ones describing emotions, several existing ontologies were found in the scientific literature, for the last two categories no appropriate ontology was identified. Therefore, as an initial step, a Twitter ontology modelling the relations between users, tweets and their associated properties had to be created. Afterwards, an aspect sentiment and emotion analysis ontology, named TweetOntoSense, which connects the expressed sentiments and emotions, the twitter messages and the analyzed entities and their facets was defined.

The main concepts from the three ontologies are shown in Fig. 3, together with the object properties that connect them. The following subsections describe in further details the proposed ontologies, used to enable social media analysis using semantic web technologies.

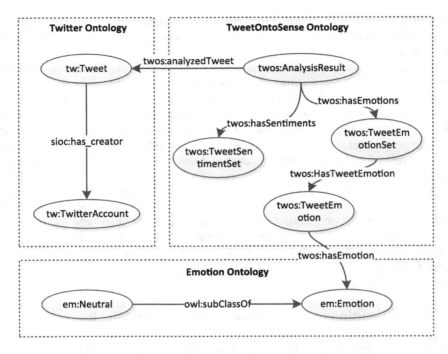

Fig. 3. Ontology-based aspect sentiment and emotion analysis

3.1 Emotion Ontology

Several emotion ontologies such as the ones proposed in [19, 24, 25] currently exist. From them, it has been chosen the emotional categories ontology presented in [25], as besides being inspired by recognized psychological models, it also structures the different human emotions in a taxonomy. The nine top-level emotions in the ontology, as well as the second-level emotions associated with the concept of "Anger", are shown in Fig. 4.

The ontology contains for each class a number of individuals, representing words associated with the particular type of emotion. In order to obtain a better coverage of the words used to express emotions, we have chosen to enrich the ontology using some of the values in the corresponding WordNet synsets [26]. Figure 5, shows the WordNet synset for the word "fear", corresponding to the concept of "Fear" in the emotion categories ontology.

Even though the ontology currently supports only English and Spanish, it can easily be extended with other languages as shown in [27], where the ontology was extended to include concepts in Italian. Thus, tweets in other languages can be more precisely analyzed, without having to resort to automatic translation services. This can prove highly important in many situations, as almost 49 % of all the Twitter messages are written in other languages than English.

The *em* prefix is used in the rest of the paper to denote classes or properties belonging to this ontology.

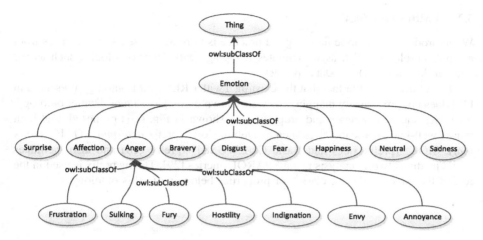

Fig. 4. Overview of the Emotion Ontology [15]

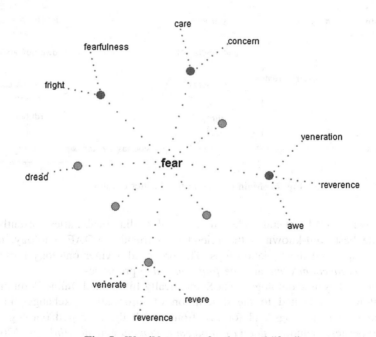

Fig. 5. WordNet synset for the word "fear"

3.2 Twitter Ontology

When producing semantic data, a good practice is to reuse classes and properties from existing ontologies [17], as it facilitates mappings with other ontologies such as the ones in the Linked Open Data[2] project.

Therefore, given the fact that the existing Twitter REST API ontology presented in [28] does not provide any mappings to well-known ontologies, a new Twitter ontology, for which the main classes and properties are shown in Fig. 6, is proposed, that both reuses well-known vocabularies such as Dublin Core[3] (prefix *dcterms*), FOAF[4] (prefix *foaf*), SIOC[5] (prefix *sioc*) and Basic Geo WGS84[6] (prefix *geo*) and also facilitates social media network analysis using SPARQL queries [29]. The *tw* prefix is used in the rest of the paper to denote classes or properties belonging to this ontology.

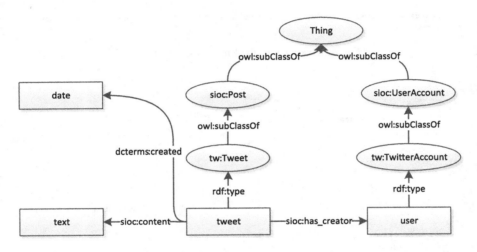

Fig. 6. Main classes in the Twitter ontology

As shown in [30], several widely used social media vocabularies currently exist. One of the best well-known is the Friend of a friend – FOAF ontology, used to represent people and their relationships. The proposed Twitter ontology reuses from FOAF the *foaf:accountName* and the *foaf:homepage* properties.

Another widely used ontology is The Semantically-Interlinked Online Communities – SIOC ontology, dedicated to the description of information exchanges in online communities such as blogs and forums, from which the proposed ontology reuses several properties, including *sioc:has_topic*, *sioc:content* and *sioc:links_to*. Moreover,

---

[2] http://linkeddata.org.

[3] http://dublincore.org/.

[4] http://xmlns.com/foaf/spec/.

[5] http://sioc-project.org/.

[6] http://www.w3.org/2003/01/geo/.

the *tw:Tweet* and *tw:TwitterAccount* classes are derived from the *sioc:Post* and *sioc: UserAccount* classes, defined in the SIOC ontology.

The Dublin Core ontology provides terms to declare a large variety of document's metadata, from which the *dcterms:created* and *dcterms:language* properties have been reused, in order to specify the date when the tweet was published and the language of the tweet.

The Basic Geo WGS84 vocabulary provides the necessary properties for describing the location associated with a tweet, through *geo:lat* and *geo:long*.

The information associated with a tweet can thus be represented as follows.

```
<http://twitter.com/13006812/status/454515103182774272>
  rdf:type tw:Tweet, owl:NamedIndividual ;
  dc:created "2014-04-11T07:03:39Z"^^xsd:dateTime ;
  tw:hasFavoriteCount 3 ;
  geo:long "6.79" ;
  geo:lat "47.52" ;
  sioc:has_creator https://twitter.com/twitterAccount1 ;
  sioc:content "tweet content" ;
  sioc:has_topic "hashtag" ;
  dcterms:language [ rdf:value "eng"^^dcterms:RFC4646 ] .
```

3.3   TweetOntoSense Ontology

The application specific ontology, shown in Fig. 7, describes the analyzed entities, like products, services or events, together with their facets and the detected emotions. The *twos* prefix is used in the rest of the paper to denote classes or properties belonging to this ontology.

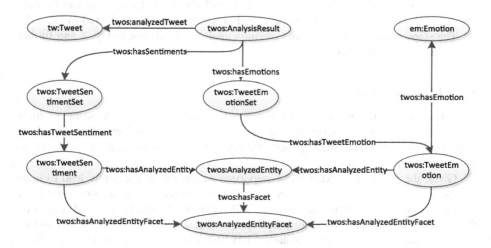

Fig. 7. Main classes in the TweetOntoSense ontology

The main classes, around which the ontology is built are *twos:AnalysisResult, twos:AnalyzedEntity* and *twos:AnalizedEntityFacet*. The *twos:Entity* class serves as a base class for *twos:AnalyzedEntity* and *twos:AnalizedEntityFacet* and defines the *twos:hasQueryTerm* data property, containing the keywords or hashtags that will be used to retrieve the analyzed tweets. The analyzed entity is modeled by the *twos:AnalyzedEntity* class, representing the particular product, service or event for which social media analysis is performed. An alternative approach for modelling the analyzed entities is presented in [31].

Given the fact that people usually express opinions not only about the concept, but also about its characteristics, known as facets [3], the *twos:AnalizedEntityFacet* class models the relevant characteristics. Finally, the *twos:AnalysisResult* class provides the necessary link with the Twitter ontology, previously described. It includes the sentiment analysis results, represented by *twos:TweetSetimentSet* and the emotion analysis result, represented by *twos:TweetEmotionSet*. The *twos:TweetSentiment* class is used to represent the detected sentiment and stores the associated strength through the *twos:hasSentimentStrength*. The *twos:TweetEmotion* class provides the link with the detected emotion from the Emotion Ontology and stores the strength of the detected emotion through the *twos:hasEmotionStrength*.

4   Ontology Based Sentiment and Emotion Analysis

The section shows how the proposed ontologies can be used to perform automatic semantic web-based sentiment and emotion social media analysis.

Extracting sentiments and emotions from tweets is known to be a challenging task for several reasons. Among the difficulties that were encountered while performing aspect sentiment and emotion analysis, it can be mentioned the huge variety of topics covered, the informality of the language, as well as the extensive usage of abbreviations and emoticons. Besides this, the concise nature of the Twitter messages can be considered both an advantage and a drawback. Further reasons are explained in [32, 33].

The steps used for sentiment and emotion analysis are shown in Fig. 8 and further described in the subsections below.

4.1   Tweet Retrieval

First, the tweets are retrieved using the Twitter Public Stream API, using as track parameters all the combinations between the keywords associated with the individuals belonging to *twos:AnalyzedEntity* and the corresponding individuals from the *twos:AnalyzedEntityFacet* class.

Given the fact that unexpected or important events can immediately lead to huge number of tweets being written every minute, the retrieved tweets are first stored in a high performance non-relational database and are only afterwards analyzed. Based on the in-depth comparison of existing non-relational databases provided in [34] and on our preliminary tests, Apache Cassandra[7] has been chosen for the proposed platform.

---

[7] http://cassandra.apache.org/.

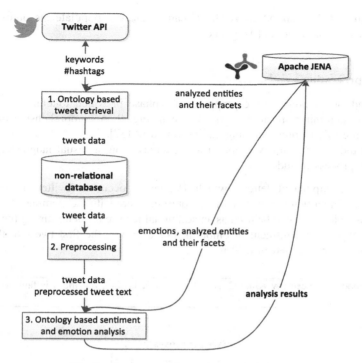

Fig. 8. Sentiment and emotion analysis steps

4.2 Language Identification

An accurate identification of the language used to write the tweet is highly important given the fact that many natural language processing algorithms and linguistic resources can only be used with the language for which they were created, with additional customizations being required for other languages.

Previously, the language had to be determined using language detection algorithms adapted for social media, such as the one presented in [35], which includes a modified version of the original TextCat identification algorithm described in [36]. The algorithm uses n-gram frequency models to discriminate between the different languages. Currently, the response received from the Twitter API also includes a field with the detected language.

While adapting the required algorithms and linguistic resources for each language, holds the promise of providing more accurate results, automatic translation, such as the one provided by Google Translate API can also be used for translating the text of the tweets written in other languages.

The Twitter and TweetOntoSense ontologies presented in this paper are language independent. However, the Emotion Ontology currently includes emotion words only for English and Spanish. MultiWordNet[8] can be used to populate the ontology with emotion words for additional languages.

4.3 Preprocessing

The second step represents the preprocessing phase in which tokenization, normalization and stemming are applied, as shown in Fig. 9. A comprehensive discussion regarding the role of preprocessing can be found in [37].

Given the fact that many users write messages using a casual language, the normalization process includes:

- Removing duplicated letters, which frequently occur in twitter messages and emphasize a particular word, in order not to interfere with the stemmer. Words with duplicated letters could be used as an additional feature for determining the intensity of the expressed sentiment or emotion. For example, the first tweet in the Sentiment140 corpus[9], used in [33] is:

"I loooooooovvvvvveee my Kindle2. Not that the DX is cool, but the 2 is fantastic in its own right."

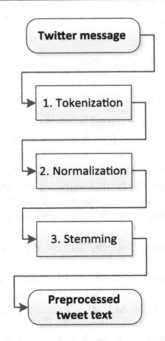

Fig. 9. Preprocessing steps

---

8 http://multiwordnet.fbk.eu/.

9 http://www.sentiment140.com/.

- Converting all-caps words to lower case. Further information regarding the intensity of the expressed sentiment or emotion can be determined based on the usage of all-caps as shown in [19]. An example of a tweet that uses All-caps for certain words is:

"My Kindle2 came and I LOVE it! :)".

- Replacing hashtags with the corresponding words.
- Replacing abbreviations with the corresponding regular words taken from the Internet Lingo Dictionary.

AHH YES LOL IMA TELL MY HUBBY TO GO GET ME SUM MCDONALDS

- Replacing emoticons with the corresponding emotions from the previously described Emotion Ontology. The Internet Lingo Dictionary[10] has been used to gather the emoticons, together with their meaning, although other sources such as the Smiley Ontology[11] could prove equally useful. A similar set of emoticons is used in [33], with the mention that they are only divided into emoticons for expressing positive and negative feelings.

Table 1 includes the emoticons that are mapped to the word happiness during the preprocessing phase.

Table 1. Emotions mapped to happiness

:)	:)	:-)	:-))	:-)))	;)	;-)	^_^	:-D	:D
=D	C:	=)							

Table 2 includes the emoticons that are mapped to the word surprise during the preprocessing phase.

Table 2. Emotions mapped to surprise

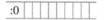

:0								

Table 3 includes the emoticons that are mapped to the word sadness during the preprocessing phase.

Table 3. Emotions mapped to sadness

:-(	:(	:((	: (	D:	Dx	'n'	:\	/:	):-/
:'	='[	:_(	/T_T	TOT	;_;	(:-(			

The last operation of the preprocessing phase consists in applying the Porter stemmer on the resulting sequence of words.

---

[10] http://www.netlingo.com/smileys.php.

[11] http://www.smileyontology.com/.

4.4   Sentiment and Emotion Identification

In the last step, sentiments and emotions are extracted from the preprocessed tweets. The proposed ontologies and approach can easily be used with adapted versions of advanced aspect-based sentiment and emotion mining algorithms, like the ones presented in [14–16].

As the novelty of the proposed approach lies in the ontology-based analysis of tweets preceding and following the sentiment analysis phase, we have chosen a simple sentiment and emotion mining approach, which only focuses on extracting explicit sentiments and emotions.

Thus, emotions are determined by comparing the processed tweet with the stemmed versions of the individuals in the enriched ontology of emotion categories. Sentiments were determined by grouping the emotions in positive, negative and neutral ones. The resulting knowledge is saved in the triple store for further analysis using SPARQL queries, as it is shown in the Sect. 5 of the paper. Even though the proposed emotion analysis approach is relatively simple, it has been found to provide fairly good results when tested on a publically available corpus[12], containing 5513 tweets collected for the search terms "Microsoft", "Apple", "Twitter" and "Google", which were annotated with the following sentiment labels: positive, negative, neutral and irrelevant. From the above mentioned corpus, only the positive and negative tweets were analyzed, as they are the ones that could express emotions. Thus, a subset of 973 tweets was selected for further analysis, representing 17.64 % from the initial set.

Table 4. Emotion analysis results on the analyzed corpus

	Apple	Google	Microsoft	Twitter	Total
Like	13	5	7	9	35
Love	7	6	6	6	25
Hate	6	1	2	1	10
Hope	3	0	1	0	4
Upset	1	0	0	0	1
	31	12	16	16	75

After comparing these tweets with the words included in the Emotion Ontology, 75 tweets were found to express emotions, the most frequent ones being "like" (35), "love" (25), "hate" (10) and "hope" (4). An overview of the results is given in Table 4. The results grouped by the topmost emotions in the ontology, are shown in Fig. 10. Analyzing the emotions at different levels of granularity can easily be performed thanks to the hierarchical organization of the various emotions in the Emotion Ontology, as it is shown in the Sect. 5 of the paper.

---

[12] http://www.sananalytics.com/lab/twitter-sentiment/.

Unrevealing a significant number of emotions even when using a simple detection approach, like the one described above, proves once more that users frequently express opinions in social media messages.

Aspect opinion mining can be performed by comparing the preprocessed tweets with the *twos:AnalizedEntityFacet* individuals associated with the *twos:AnalyzedEntity* for which the tweet has been retrieved.

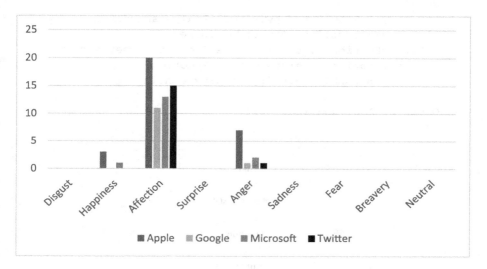

Fig. 10. Emotion analysis results on the analyzed corpus

5   Social Media Analysis

The proposed semantic analysis approach can be used to develop an end-to-end ontology-based social media analysis platform. A complete semantic approach is provided through:

- modelling the analyzed emotions using the selected Emotion Ontology;
- modelling the Twitter related data using the proposed Twitter Ontology;
- modelling the analyzed entities, representing for example products or services, their facets and the analysis results using the proposed TweetOntoSense Ontology.

The approach offers multiple advantages, including the possibility to exploit the vast amount of information readily available in the Linking Data Cloud using the technologies associated with the semantic web.

Moreover, using semantic web inference, for example, new relations between the collected information can be discovered automatically. Thus, if the *em:offended* emotion is associated to a tweet, during the emotion identification phase, the inference engine also associates the more general *em:indignation* and *em:anger* emotions. Figure 11 shows the hierarchy relation between the three emotions in the Emotion Ontology. Therefore, emotion analysis can easily be performed at various granularity levels.

SPARQL queries provide the necessary mean for performing advanced analysis, while their structured result can easily be processed for creating meaningful visualizations at the user interface level. For example, the following query retrieves from the triple store all the studied entities together with the detected emotions.

```
SELECT ?analyzedEntity ?emotion
WHERE
{ ?analysisResult rdf:type twos:AnalysisResult;
    twos:hasEmotions ?tweetEmotionSet.
  ?tweetEmotionSet twos:hasTweetEmotion ?tweetEmotion.
  ?tweetEmotion twos:hasEmotion ?emotion;
    twos:hasAnalyzedEntity ?analyzedEntity.
}
ORDER BY ASC(?analizedEntity)
```

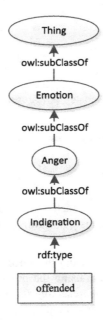

Fig. 11. Emotion hierarchy

Using also inference, the query bellow returns the users that have written tweets which express emotions derived from *em:happiness*, ordered by the influence of each user, measured as the number of followers. The number of followers that a user has, indicates the potential reach of the social media messages that he or she posts. Users with a large follower base, are called influencers and their positive or negative opinions can have an important effect on a large number of people. Therefore, identifying and engaging them is considered a priority by social media practitioners. Using the approach proposed in this paper, influencers can easily be determined for each category or sub-category of emotions.

```
SELECT ?user ?tweetContent ?followerCount
WHERE
{
    ?analysisResult rdf:type twos:AnalysisResult;
       twos:hasEmotions ?tweetEmotionSet;
       twos:analyzedTweet ?tweet.
    ?tweetEmotionSet twos:hasTweetEmotion ?tweetEmotion.
    ?tweetEmotion twos:hasEmotion ?emotion.
    ?tweet sioc:content ?tweetContent;
       sioc:hasCreator ?user.
    ?user tw:hasFollowerCount ?followerCount.
    ?emotion rdf:type em:Happiness.
}
ORDER BY DESC(?followerCount)
```

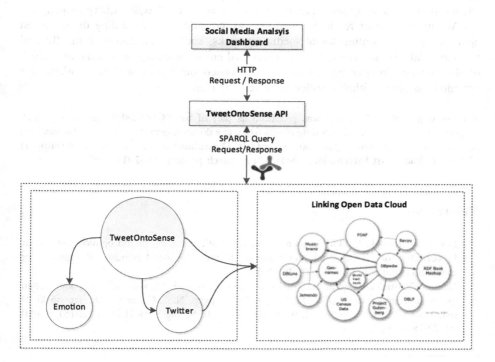

Fig. 12. Social media analysis platform

The architecture of a semantic social media platform, using the proposed approach is shown in Fig. 11. The social media analysis dashboard, representing the user interface of the platform, can communicate with a common REST approach with a Web API, labeled TweetOntoSense API in the figure. The API performs the necessary SPARQL queries on the semantic database and returns the results to the social media analysis dashboard in JavaScript Object Notation – JSON format.

By reusing classes from the Linking Open Data Cloud, advanced analysis can be performed, that tap into the vast information available in knowledgebase such as DBpedia[13] (Fig. 12).

6   Concluding Remarks

The present paper proposes a novel ontology-based social media sentiment and emotion analysis approach that better captures the wide array of feelings expressed in the millions of tweets published every day. By using a multiple level emotion ontology, the presented approach paves the way towards fine-grained emotion analysis using semantic web technologies, thus unlocking a vast amount of emotional information that has previously been unavailable to companies and public authorities, trying to better understand their customers' opinions through social media analysis. By using concepts instead of keywords and storing the social media analysis results as structured data described with the help of ontologies, the proposed approach creates the bases for advanced analysis using semantic web inference and the SPAQRL query language.

Among the further research directions, we consider both extending the proposed approach to other online social media networks, such as Facebook, LinkedIn and Google+ and also analyzing how the expressed emotions change over time as a result of the changes in user perception. The proposed ontologies will be available for download at https://github.com/lcotfas/TweetOntoSense.

Acknowledgments. The study was produced as part of the SCOPANUM research project, supported by grants from CSFRS (http://csfrs.fr/), and a doctoral grant from Pays de Montbéliard Agglomération (http://www.agglo-montbeliard.fr/). **The authors also acknowledge the support of Leverhulme Trust International Network research project "IN-2014-020".**

References

1. Pak, A., Paroubek, P.: Twitter as a corpus for sentiment analysis and opinion mining. In: Proceedings of the Seventh International Conference on Language Resources and Evaluation, Valletta, pp. 1320–1326 (2010)
2. Delcea, C., Bradea, I., Paun, R., Scarlat, E.: How impressionable are you? - grey knowledge, groups and strategies in OSN. In: Núñez, M., Nguyen, N.T., Camacho, D., Trawiński, B. (eds.) ICCCI 2015. LNCS, vol. 9329, pp. 171–180. Springer, Heidelberg (2015). doi:10. 1007/978-3-319-24069-5_16

---

[13] http://wiki.dbpedia.org/.

3. Kontopoulos, E., Berberidis, C., Dergiades, T., Bassiliades, N.: Ontology-based sentiment analysis of Twitter posts. Expert Syst. Appl. **40**, 4065–4074 (2013)
4. Delcea, C., Cotfas, L.-A., Paun, R.: Understanding online social networks' users – a Twitter approach. In: Hwang, D., Jung, J.J., Nguyen, N.-T. (eds.) ICCCI 2014. LNCS, vol. 8733, pp. 145–153. Springer, Heidelberg (2014)
5. Delcea, C., Cotfas, L.-A., Paun, R.: Grey social networks. In: Hwang, D., Jung, J.J., Nguyen, N.-T. (eds.) ICCCI 2014. LNCS, vol. 8733, pp. 125–134. Springer, Heidelberg (2014)
6. Torkildson, M.K., Starbird, K., Aragon, C.: Analysis and visualization of sentiment and emotion on crisis tweets. In: Luo, Y. (ed.) CDVE 2014. LNCS, vol. 8683, pp. 64–67. Springer, Heidelberg (2014)
7. Rill, S., Reinel, D., Scheidt, J., Zicari, R.V.: PoliTwi: Early detection of emerging political topics on Twitter and the impact on concept-level sentiment analysis. Knowl.-Based Syst. **69**, 24–33 (2014)
8. Delcea, C., Bradea, I., Paun, R., Friptu, A.: A Healthcare companies' performance view through OSN. In: Barbucha, D., Nguyen, N.T., Batubara, J. (eds.) New Trends in Intelligent Information and Database Systems. SCI, vol. 598, pp. 333–342. Springer, Heidelberg (2015)
9. Nassirtoussi, A.K., Aghabozorgi, S., Wah, T.Y., Ngo, D.C.L.: Text mining for market prediction: a systematic review. Expert Syst. Appl. **41**, 7653–7670 (2014)
10. Ghiassi, M., Skinner, J., Zimbra, D.: Twitter brand sentiment analysis: a hybrid system using n-gram analysis and dynamic artificial neural network. Expert Syst. Appl. **40**, 6266–6282 (2013)
11. Mostafa, M.M.: More than words: social networks' text mining for consumer brand sentiments. Expert Syst. Appl. **40**, 4241–4251 (2013)
12. Thelwall, M., Buckley, K., Paltoglou, G.: Sentiment strength detection for the social web. J. Am. Soc. Inf. Sci. Technol. **63**, 163–173 (2012)
13. Robaldo, L., Di Caro, L.: OpinionMining-ML. Comput. Stand. Interfaces **35**, 454–469 (2013)
14. Medhat, W., Hassan, A., Korashy, H.: Sentiment analysis algorithms and applications: a survey. Ain Shams Eng. J. **5**, 1093–1113 (2014)
15. Tsytsarau, M., Palpanas, T.: Survey on mining subjective data on the web. Data Min. Knowl. Discov. **24**, 478–514 (2012)
16. Liu, B.: Sentiment analysis and opinion mining. Synth. Lect. Hum. Lang. Technol. **5**, 1–167 (2012)
17. Shadbolt, N., Berners-Lee, T., Hall, W.: The semantic web revisited. IEEE Intell. Syst. **21**, 96–101 (2006)
18. Borst, W.N.: Construction of engineering ontologies for knowledge sharing and reuse. Universiteit Twente (1997)
19. Roberts, K., Roach, M., Johnson, J.: EmpaTweet: annotating and detecting emotions on Twitter. In: Proceedings of the Eighth International Conference on Language Resources and Evaluation, Istanbul, pp. 3806–3813 (2012)
20. Baggia, P., Burkhardt, F., Pelachaud, C., Peter, C., Zovato, E.: Emotion Markup Language (EmotionML) 1.0. http://www.w3.org/TR/emotionml/
21. Ashimura, K., Baggia, P., Burkhardt, F., Oltramari, A., Peter, C., Zovato, E.: Vocabularies for EmotionML. http://www.w3.org/TR/2012/NOTE-emotion-voc-20120510/
22. Westerski, A., Iglesias Fernandez, C.A., Tapia Rico, F.: Linked opinions: describing sentiments on the structured web of data. Presented at the 4th International Workshop Social Data on the Web, Bonn, Germany (2011)
23. Belhajjame, K., Cheney, J., Corsar, D., Garijo, D., Soiland-Reyes, S., Zednik, S., Zhao, J.: PROV-O: The PROV Ontology. http://www.w3.org/TR/prov-o/

24. Hastings, J., Ceusters, W., Smith, B., Mulligan, K.: Dispositions and processes in the Emotion Ontology. In: Proceedings of the ICBO 2011 (2011)
25. Francisco, V., Hervás, R., Peinado, F., Gervás, P.: EmoTales: creating a corpus of folk tales with emotional annotations. Lang. Resour. Eval. **46**, 341–381 (2012)
26. Montejo-Ráez, A., Martínez-Cámara, E., Martín-Valdivia, M.T., Ureña-López, L.A.: Ranked WordNet graph for sentiment polarity classification in Twitter. Comput. Speech Lang. **28**, 93–107 (2014)
27. Baldoni, M., Baroglio, C., Patti, V., Rena, P.: From tags to emotions: ontology-driven sentiment analysis in the social semantic web. Presented at the Intelligenza Artificiale (2012)
28. Togias, K., Kameas, A.: An ontology-based representation of the Twitter REST API. Presented, November 2012
29. Cotfas, L.-A., Delcea, C., Roxin, I., Paun, R.: Twitter ontology-driven sentiment analysis. In: Barbucha, D., Nguyen, N.T., Batubara, J. (eds.) New Trends in Intelligent Information and Database Systems. SCI, vol. 598, pp. 131–139. Springer, Cham (2015)
30. Breslin, J.G., Passant, A., Decker, S.: The Social Semantic Web. Springer, Heidelberg (2009)
31. Fornara, N., Ježić, G., Kušek, M., Lovrek, I., Podobnik, V., Tržec, K.: Semantics in multi-agent systems. In: Ossowski, S. (ed.) Agreement Technologies, pp. 115–136. Springer, Dordrecht (2013)
32. Maynard, D., Bontcheva, K., Rout, D.: Challenges in developing opinion mining tools for social media. In: Proceedings of the Eighth International Conference on Language Resources and Evaluation, Istanbul, pp. 15–22 (2012)
33. Kouloumpis, E., Wilson, T., Moore, J.: Twitter sentiment analysis: the good the bad and the omg! In: Proceedings of the Fifth International Conference on Weblogs and Social Media, Barcelona, pp. 538–541 (2011)
34. Rabl, T., Gómez-Villamor, S., Sadoghi, M., Muntés-Mulero, V., Jacobsen, H.-A., Mankov-skii, S.: Solving big data challenges for enterprise application performance management. Proc. VLDB Endowment **5**, 1724–1735 (2012)
35. Carter, S., Weerkamp, W., Tsagkias, M.: Microblog language identification: overcoming the limitations of short, unedited and idiomatic text. Lang. Resour. Eval. **47**, 195–215 (2013)
36. Cavnar, W., Trenkle, J.: N-gram-based text categorization. In: Proceedings of the 3rd Annual Symposium on Document Analysis and Information Retrieval, SDAIR 1994 (1994)
37. Bao, Y., Quan, C., Wang, L., Ren, F.: The role of pre-processing in Twitter sentiment analysis. In: Huang, D.-S., Jo, K.-H., Wang, L. (eds.) ICIC 2014. LNCS, vol. 8589, pp. 615–624. Springer, Heidelberg (2014)

Web Projects Evaluation Using the Method of Significant Website Assessment Criteria Detection

Paweł Ziemba[1(✉)], Jarosław Jankowski[2], Jarosław Wątróbski[2], and Mateusz Piwowarski[2]

[1] The Jacob of Paradyż University of Applied Sciences in Gorzów Wielkopolski, Teatralna 25, 66-400 Gorzów Wielkopolski, Poland
pziemba@pwsz.pl
[2] Faculty of Computer Science and Information Technology, West Pomeranian University of Technology, Szczecin, Żołnierska 49, 71-210 Szczecin, Poland
{jjankowski,jwatrobski,mpiwowarski}@wi.zut.edu.pl

Abstract. The research presented in the article consists of an examination of the applicability of feature selection methods in the task of selecting website assessment criteria, to which weights are assigned. The applicability of the chosen methods was examined against the approach in which the weightings of website assessment criteria are defined by users. The research shows a selection procedure concerning significant choice criteria and reveals undisclosed user preferences based on the website quality assessment models. Results concerning undisclosed preferences were verified through a comparison with those declared by website users.

Keywords: Website evaluation quality · User experience · Feature selection

1 Introduction

In today's society, which relies heavily on information, with most official matters and commercial transactions being conducted through the Internet, the quality and usability of a website is highly significant to the users' perception of an organization. This can be observed in the case of e-commerce, e-government and many other services. While almost 634 million web pages operate in the world[1] with more than 2.4 billion users[2], the quality of a website has also a major impact on the number of users visiting the site. Assessment of a website's quality is therefore extremely important. Some systems have a global reach; others are local. News portals are one of the most popular types of websites, with their popularity being strictly linked with quality. This hypothesis is present in publications concerning website assessment, whereby customers must be satisfied with their experience in using the website or they will not return. Thus, the assessment of website quality has become a priority for companies [1] which affects

---

[1] http://news.netcraft.com/archives/2012/12/04/december-2012-web-server-survey.html.
[2] http://www.internetworldstats.com/stats.htm.

© Springer-Verlag Berlin Heidelberg 2016
N.T. Nguyen and R. Kowalczyk (Eds.): TCCI XXII, LNCS 9655, pp. 167–188, 2016.
DOI: 10.1007/978-3-662-49619-0_9

users' loyalty and usage frequency [2]. The importance of website quality assessment is reflected in other research which states that the effective evaluation of websites has become a point of concern for practitioners and researchers [3]. Evaluation is an aspect of website development and operation that can contribute to maximizing the exploitation of invested resources [4]. The assessment of website quality, including the most popular types of websites (i.e. news portals and social platforms), is a challenging task in most stages of online projects. An effort should be made to ensure that the website is reliable and reflects users' expectations. This fact is supported in practice by the ongoing development of methods of presenting information as well as information accessibility and the development of new functions tailored to meet the current needs of users.

There are various models of website quality assessment that can be found in the literature, including: eQual [5], Ahn [6], SiteQual [7], Web Portal Site Quality [8] and Website Evaluation Questionnaire [9]. One of the limitations of these models is the method of obtaining criteria weights during the evaluation of services, as they are usually defined on the basis of a declarative approach. Meanwhile, the criteria for determining weights based on surveys and explicit declared user preferences can generate errors in the study [10]. Declared user preferences may differ from actual preferences and from preferences acquired by analytical systems [11]. In addition, their permanent use in assessing the quality of all the criteria derived from these models would be very confusing, because there is a total of more than sixty of them.

The aim of this paper is to formulate a procedure for selecting the significant website assessment criteria and assigning them weights, derived from data mining techniques and machine learning algorithms. This procedure is designed to formalize the process of criteria selection for website quality assessment methods. It should simplify the process of obtaining website assessments by reducing the number of criteria contained in user surveys and expert surveys. A reduced set of criteria, obtained by applying the developed procedure, should give results close to the real users' reviews. Moreover, the procedure should allow for the identification undisclosed user preferences, which in reality are favored when assessing the website, and which may differ from those explicitly declared. In combination with procedures for obtaining partial opinions and determining the final assessment of websites, this procedure may become a part of the expert system for determining the quality of websites.

Proposed in this paper approach is an extended version of the approach shown in [12] based on determining the criteria that implicitly influence users in their evaluation of the quality of websites. This procedure allows the usage of weights that reflect their relevance and enable criteria reduction with a low impact on the assessment.

The article subsequently presents approaches to assessing the quality of web services and the feature selection techniques used in machine learning applied in the approach proposed by the authors. Then the procedure of criteria selection for assessing the quality of websites is presented along with the results, followed by a discussion of the resulting solution. The article ends with conclusions and provides recommendations for further directions of work.

2 Literature Review and Proposed Approach

2.1 Website Quality Evaluation Models and Methods for Obtaining Evaluations

The quality of a website can be understood as the qualification of how well it meets the needs of users [13]. It should be noted that quality is defined by a model composed of characteristics and features/criteria describing its various components [14]. Various models of website assessment, varying in their uses of quality description criteria, the number of criteria, assessment scales and methods require a model which focuses on the information value of the researched sites. According to the analysis, the following models have the highest usability in assessment of the news portals: eQual [5], Ahn [6], SiteQual [7], Web Portal Site Quality [8] and Website Evaluation Questionnaire [9]. The statement was influenced by the fact that these models elaborately treat the issue of the quality of the presented information. In addition, they have been widely used in both academic work [15] and business practice [16]. The main problem with the quality assessment models refers to the method for obtaining evaluations. This can be achieved through an expert evaluation, surveys or user activity tracking.

User activity tracking can be done by eye or mouse tracking [17]. Eye tracking is very expensive in terms of time, finances and calculations. This method requires specialized equipment or software. This method of data collection is time-consuming, which results in a very small research sample, and processing this kind of data requires complex calculations [18]. Additionally, the applicability of eye or mouse movement tracking is limited to an evaluation of websites' usability [19], rather than their quality.

At the same time usability and quality are two concepts which are close to each other, but which cover a diverse range of meaning. Nielsen concluded that usability is the attribute of the quality describing how easy the user interface is to use [20]. Based on this definition, it can be assumed that usability is just one element contributing to the quality. This is confirmed by ISO 9126 and ISO 25010 norms, in which the usability is treated as one of the components of the quality of [14, 21]. Also, in many models the research and quality evaluation of websites' usability is mentioned as one of the components of quality [22, 23]. Therefore, the methods used to evaluate the usability of services may include too few aspects of the website with regard to its quality.

Expert and survey methods use a large number of criteria, so they are costly in time. In addition, respondents and experts often complain about the large number of evaluation criteria, which can cause carelessness in the assessment and reduce the correctness of the evaluation. A similar problem concerns the methodology of creating the website quality assessment models. When creating new assessment methods, the authors often use existing quality assessment models. Authors choose assessment criteria that, according to them, seem to be suitable for their new method. The selection process is usually performed by the authors in an informal way, based on the literature analysis and their own opinions [24, 25]. This expert approach is rarely based on clear methodologies and analysis is usually chaotic however attempts are observed to apply quantitative approach [11]. Additionally, there is a risk that the models created in this way will operate criteria of low real importance, not relevant for the website users. It can be described as the curse of dimensionality, based on the problem of choosing from many quality features only those that are useful in assessing the quality of a specific type of website. For example, criteria that are significant in evaluating news portals may not be important in assessing e-commerce websites. Some authors use a more formalized approach to the selection of criteria; based on surveys, the criteria are grouped by a statistical method, such as the factor analysis. They eliminate those criteria that, according to the results of the analysis, do not belong to any of the obtained groups of criteria, or to any of the dimensions of quality [8, 26–28]. New classes of criteria are obtained through the application of factor analysis, containing subgroups of the original criteria. This approach is much more appropriate than the last, but there is also a problem here concerning the assessment of weights for each criterion. Due to the rotations and scaling of coordinates, caused by the application of factor analysis (the principal component analysis), it is difficult to determine which criteria are important. These transformations can also cause major changes in the results [29]. In this approach, the criteria weights are determined by the survey results, as averaged values of the weights given by the respondents or experts. Obtaining the criteria weights based on surveys (i.e. explicitly declared user preferences) can generate large errors [30]. This is confirmed by research demonstrating the fact that the explicitly declared preferences of users may differ from the criteria that are really used by users in their website evaluation [31, 32]. In this paper it is proposed to solve the mentioned problems concerning the selection criteria for assessing the quality of web services and taking into account weights criteria through surveys with the use of data mining techniques and machine learning algorithms.

2.2 Feature Selection Methods Used in Machine Learning

Data mining techniques are used to extract patterns from data sets that, due to the scope of the analyzed data, are not recognized by people. Data mining methods are mainly based on machine learning algorithms. Machine learning tasks concentrate on predicting an object's value or class affiliation; based on its features the multidimensionality of an object, which is to be classified into a specific category, creates a problem for classification techniques, as well as for all data mining methods. This problem is referred to as the curse of dimensionality [33]. A reduction in the number of dimensions undergoing classification allows for a reduction of calculation demands and data collection demands, as well as the increased reliability of predicate results and data quality [34].

The reduction of dimensions can be conducted with the help of two methods. The first is based on the application of a feature extraction process, which relates to the extraction of a set of new characteristics from the original features. This process usually involves remapping the original features in a way that creates new variables. Factor analysis is an example of this type of dimension reduction. The second method assumes the use of a feature selection process, concentrated on pinpointing significant features within the data set and rejecting redundant attributes. Various evaluation algorithms are used to assess the features according to a specific criterion, describing their significance for the classification task [35].

The feature selection process can be regarded as searching through a set of characteristics describing an object undergoing classification, according to a particular assessment criterion. The process entails three procedures: filters, wrappers and embedded methods. Filters are based on independent feature assessments, using general data characteristics. Feature sets are filtered in order to establish the most promising attribute subset before commencing machine learning algorithm training [36]. Wrapper functions evaluate specific feature subsets with the help of machine learning algorithms. The learning algorithm is, in this case, included in the feature selection procedure [37]. In embedded methods, as in the wrappers, feature selection is linked to the classification stage. In other words, embedded methods use the internal information of the classification model to perform feature selection [38]. Subsets of features selected by a filter or a wrapper are further evaluated using the machine learning algorithm [39].

Wrapper functions differ from one another only in terms of utilized machine learning algorithms, so the results obtained with the help of wrappers depend only on the quality of the machine learning algorithm and whether or not it suits the given classification task. Wrappers and embedded methods analyze the features of objects only in terms of obtaining the maximum level of correct classification, ignoring other characteristics of the features. In the examined procedure, the feature selection method used in data mining is only one of the stages of the entire algorithm. Obtaining the best results of classification is not the main objective. Furthermore, the general characteristics of the criteria seem to be important enough to influence the choice of evaluation criteria. Therefore, the use of filters that are based on the general characteristics of features, rather than wrappers and embedded methods, would appear appropriate in the studied procedure. The filters utilize methods which involve proximity function, probability theory, and various measures of correlation.

The most popular filter utilizing the proximity function is a ReliefF method. The primary idea behind the ReliefF method is to evaluate attributes according to how well they differentiate between similar objects, i.e. objects of similar feature values. The nearest neighbor method is used here, as another example of a proximity function [40]. The ReliefF procedure utilizes a heuristic rule according to which a good attribute should differentiate between objects situated close to each other but belonging to various classes, and it should also maintain the same value for objects situated close to each other but belonging to the same class [41].

One of the best known methods employing the probabilistic approach in order to establish the direction of the correct solution is the LVF (Las Vegas Filter) procedure. Solution searches are conducted randomly, which guarantees an acceptable solution even if incorrect decisions are made during the search for the best subset. This method uses the inconsistency criterion to determine the level of acceptance of data with reduced dimensionality [33, 42].

The correlation procedures constitute the largest group of filters. Amongst them, methods using conventional measures of correlation (such as Spearman's and Pearson's) and nonstandard measures of correlation can be distinguished. Within the unconventional correlation measures, the most interesting ones in terms of the calculation procedures are: SA (Significance of Attribute), Hellwig method (individual indicators of Hellwig information capacity), CFS (Correlation-based Feature Selection) and FCBF (Fast Correlation Based Filter). The SA method utilizes the correlation coefficient's bidirectional links between attributes and class affiliation. This method is based on a heuristic stating that if an attribute is significant, then there is a high probability that objects complementing the value sets of this attribute will belong to the complementary sets of class. Additionally, assuming that the decision classes for two sets of objects are different, it can be expected that the attribute significance value for objects belonging to two different sets will also differ [43]. The FCBF method is based on the correlation coefficient or, more precisely, symmetrical uncertainty. Additionally, as auxiliary means, the FCBF method utilizes sets of redundant features separately for each feature. The FCBF procedure initially involves calculation of the symmetrical uncertainty for each feature, and further considerations involve only attributes with symmetrical uncertainty values higher than the assumed threshold [44]. The CFS method, as well as the FCBF method, is based on the analysis of correlation between features. The global measure of correlation used by the CFS procedure is Pearson's linear correlation, whilst symmetrical uncertainty is considered a local measure. A heuristic applied for CFS states that a good feature subset contains attributes strongly correlated with a specific class of objects but uncorrelated with other classes and attributes [37, 45]. The Hellwig method is similar to CFS, since it is based on the Pearson correlation and it consists of the selection of features that are strongly correlated with the class of object, and weakly correlated with each other [46]. Due to the high computational complexity of the Hellwig method, in these studies only individual indicators of Hellwig information capacity were applied. Symmetrical uncertainty used in FCBF and CFS procedures is also a separate method of feature selection. Symmetrical uncertainty is one of the correlation coefficients specifying the information gain based on the theoretical concept of entropy. SU compensates for the information gain's bias toward features with more values and normalizes its values within the range [0, 1], with the value 1 indicating that knowledge of either one of the values completely predicts the value of the other and the value 0 indicating that X and Y are independent. It treats a pair of features symmetrically [47]. The research discussed further below examines the applicability of the feature selection methods in the task of selecting website assessment criteria and assigning weights to them. The applicability of the chosen methods was tested against the approach in which the weights of website assessment criteria are defined by the users.

2.3 Proposed Approach

The proposed procedure is characterized by the possibility of generalization for applying it to the selection criteria relevant when assessing various types of sites. The presented method for determining the selection criteria and their weights is based on the assumption that the surveyed multiple criteria evaluation is not accurate. This can be proved by conducting a survey in which respondents evaluate websites in terms of succeeding criteria. They attribute evaluations to the websites in relation to the criteria, as well as an overall assessment. The evaluation is only accurate if the websites' ratings, calculated as weighted averages of assessment criteria, correspond with the overall evaluations of these websites. The second assumption states that there is a subset of the criteria which differentiates website quality to a considerable degree. There are also implicit values of weights to be used in multi-criteria assessment services which can give a solution similar to the overall assessment of service determined by the respondents. Based on the assumptions formulated in the research, heuristics inspired by algorithms used in the construction of feature selection methods of machine learning were applied. According to this approach for finding the subset of classification criteria, there is a model in the sense of machine learning - characterized by a low degree of conflict cases - training model which enables us to a large extent to determine the overall assessment of the service on the basis of the criterion ratings. Therefore, building models of classifiers using subsets of criteria that can be chosen subsets may provide solutions close to optimum.

For the purpose of this paper, the feature selection methods applied in machine learning were used in the selection of criteria subsets. The low number of conflicting cases influencing the learning model is also of great importance for the further investigation within the planned expert system area. This is due to the fact that at this stage of the research the use of correct classification rates is accepted as an additional form of the evaluations' weighting in a website assessment system.

3 Procedure for Selecting Website Evaluation Criteria

The developed procedure was presented using the model of website quality created from five source models: eQual, Ahn, SiteQual, Web Portal Site Quality and Website Evaluation Questionnaire. The decision was influenced by the fact that the models are the best-formalized models among those taken under consideration, and their numerous applications indicate their high levels of universality. Additionally, as in the case of other models, they use declared user preferences. These models are supposed to portray the quality of websites from the user's perspective. Therefore, they utilize the survey method to obtain opinions regarding the websites. The evaluators declare the degree of conformity of each criteria with their opinion concerning a website on a Likert point scale. The same scale is used to declare the significance of criteria. Criteria used by particular models are presented in Tables 1, 2, 3, 4 and 5.

Table 1. eQual model assessment criteria

No.	eQual criteria
1	I find the site easy to learn to operate
2	My interaction with the site is clear and understandable
3	I find the site easy to navigate
4	I find the site easy to use
5	The site has an attractive appearance
6	The design is appropriate to the type of site
7	The site conveys a sense of competency
8	The site creates a positive experience for me

(Continued)

Table 1. (*Continued*)

No.	eQual criteria
9	Provides accurate information
10	Provides believable information
11	Provides timely information
12	Provides relevant information
13	Provides easy-to-understand information
14	Provides information at the right level of detail
15	Presents the information in an appropriate format
16	Has a good reputation
17	It feels safe to complete a transaction
18	My personal information feels secure
19	Creates a sense of personalization
20	Conveys a sense of community
21	Makes it easy to communicate with the organization
22	I feel confident that goods/services will be delivered as promised

Table 2. Ahn model assessment criteria

No.	Ahn criteria
23	Has fast response and transaction processing
24	Can use when I want to use
25	Has good functionality relevant to site type
26	Keeps error-free transactions
27	Provides complete information
28	Provides site-specific information
29	Instils confidence in users, reducing their uncertainty
30	Provides follow-up service to users

Table 3. SiteQual model assessment criteria

No.	SiteQual criteria
31	Handles service requests dependably
32	Minimizes distractions
33	Anticipates and answers customer questions on website
34	Keeps the customer's best interests at the forefront
35	Deals with customers in a courteous manner
36	Provides latest technology
37	Each website component is designed for visual appeal
38	Site is registered for easy location (search engines)

(*Continued*)

Table 3. (*Continued*)

No.	SiteQual criteria
39	Provides for internal search capability
40	Displays the right amount of information for the task without overload
41	Provides a value-added experience
42	Uses consistent standardized representations/metaphors
43	Eliminates bias in information provided
44	Demonstrates commitment to privacy of personal information

Table 4. Website Evaluation Questionnaire model assessment criteria

No.	Website Evaluation Questionnaire criteria
45	I consider this website user-friendly
46	The homepage clearly directs me towards the information I need
47	The homepage immediately points me to the information I need
48	Under the hyperlinks, I found the information I expected to find there
49	I know where to find the information I need on this website
50	I find the structure of this website clear
51	The convenient set-up of the website helps me find the information
52	I find the information in this website helpful
53	This website offers information that I find useful
54	The language used in this website is clear to me
55	I find the design of this website appealing
56	The search option on this website helps me to find the information quickly
57	The search option on this website gives me useful results

Table 5. Web Portal Site Quality model assessment criteria

No.	Web Portal Site Quality criteria
58	Customized search functions
59	Search facilities
60	Adequacy of security features
61	Valuable tips on products/services
62	Unique content
63	Complete product/service description
64	Relatively comprehensive information compared to other portals
65	Detailed contact information
66	High speed of page loading
67	Message board forum

The aim of the proposed procedure is to analyze the data obtained using feature selection procedures. The objects are to be analyzed with sets of marks awarded by each of the respondents for each service relative to successive criteria and the final marks. Each object consists, therefore, of a set of features, which are criteria for evaluation and descriptions of the class to which the object belongs in the form of global assessment.

Feature selection methods examined the influence of individual characteristics on assigning an object to a specific class. An independent assessment of the characteristics of the use of the general characteristics of the data is carried out. Here the correlation coefficients between the values and characteristics belonging to a specific class can be used. These methods, as opposed to wrappers, choose characteristics regardless of the results of the classification, and the classifier is used only to verify the set of characteristics [39]. The use of filters eliminates the impact of the quality of the classifier to select features. In addition, the method was used to examine further features independently of each other, resulting in a ranking of the full set of features together with the numerical value of the significance of each feature. During the processes the following methods were used: ReliefF [41], Significance Attribute [43], Symmetrical Uncertainty [47], individual indicators of Hellwig information capacity [46], CFS (Correlation-based Feature Selection) [37], FCBF (Fast Correlation Based Filter) [44] and LVF (Las Vegas Filter) [33]. The ReliefF, SA, SU and Hellwig methods were used in order to examine subsequent features independently of the others. They enabled the achievement of rankings for the full sets of criteria. An output from the remaining methods (FCBF, CFS, LVF) gave reduced criteria subsets, with the significance level of the other criteria remaining at 0. In order to achieve the numerical significance of selected criteria in the case of CFS method, the Hill Climbing search strategy was used rather than the Best First. For the LVF and CFS-HC methods, the fact that they provide numerical evaluation of selected features in reverse order should be taken into account, i.e. the worst feature has the highest value in evaluation. Therefore, the significance of the criteria selected through these methods was calculated by subtracting from the value "1" the result acquired from the LVF or CFS-HC method. To evaluate the feature for the ReliefF method, 10 nearest neighbors were applied, with the sampling performed on all objects. After the rankings and subsets of features, testing should be performed using the methods of classification. A decision tree classifier – namely, the advanced classification tree CART – was deployed for this purpose. The Gini measure was used as a criterion for the distribution node in the tree. The minimum allowable cardinality of the node was set at five objects and the stop parameter for trimming the tree was misclassification error [48]. Also used was the estimation of the a priori probability of belonging to particular classes of objects [49]. Estimating a priori helped to improve the model classifier due to the fact that the frequencies of particular classes of decision-making were different. They were close to normal distribution, so the use of the a priori estimate was justified. In order to obtain stable results of the classification the 10-fold cross validation was used [50]. The classification of objects into one of seven classes represented the specific assessment of the overall service. Each of the ranking criteria was iteratively eliminated in accordance with an important feature of the test rankings and other features used for classification. On the basis of the classification, a true positive rate and Cohen's kappa were determined. Cohen's kappa coefficient describes the compatibility between the expected and predicted objects belonging to the classes of decision-making. An important advantage of Cohen's Kappa coefficient, compared to a ratio of relevance classifier, is that it corrects a random compatibility classification [51]. Moreover, in the literature linguistic interpretations of the extent of compliance can be found, specified by the numeric value of Cohen's kappa coefficient:

- $K_C \in [0.0, 0.2]$: slight;
- $K_C \in (0.2, 0.4]$: fair;
- $K_C \in (0.4, 0.6]$: moderate;
- $K_C \in (0.6, 0.8]$: substantial;
- $K_C \in (0.8, 1]$: almost perfect [52].

The next stage of the study was to search for suboptimal subsets of criteria that would get the results of the evaluation criterion as close as possible to the results of the assessment of general services. For this purpose, based on the ratings of well-known news services given by respondents for each website included in the survey, the average overall score was calculated according to the formula (1):

$$G_s = \frac{\sum_{i=1}^{n} s_i}{L_{max}} /n * 100\,\%. \tag{1}$$

s_i – overall assessment of the service assigned by the i-th user,
n – the number of users participating in the survey,
L_{max} – evaluating the maximum value of the scale (in this case, seven).

For each obtained subset of criteria the average standardized assessment services were also calculated, using criterial evaluation and weight according to the formula (2):

$$O_s = \frac{\sum_{j=1}^{m} \sum_{i=1}^{n} \left(w_j * k_{ij} \right)}{\sum_{j=1}^{m} \left(w_j * L_{max} \right)} /n * 100\,\% \tag{2}$$

k_{ij} – evaluation of service terms of j-th criterion, assigned by the i-th user,
w_j – weight of the j-th criterion,
m – number of criteria, in terms of evaluating service.

Because different numbers of criteria resulted in obtaining the services of a different number of points, the score for each subset of criteria has been normalized to the range [0–1]. If a selected subset of the assessment criteria accurately reflects the quality of Internet service, it is between the assessment and the assessment of the overall service relationship or criterion $G_S \approx O_S$. A comparison of standard G_S and O_S values obtained for the different subsets was performed using a mean absolute deviation measure of [53], according to the formula (3):

$$MAD = \frac{\sum_{i=1}^{n} \|O_i - G_i\|}{n} \tag{3}$$

n - number of respondents' websites,
G_i - average overall rating of the i-th website,
O_i - average the criterion of the i-th website.

The final selection of a subset of criteria was based on the results of the classification and co-factors, and Cohen kappa values of mean absolute deviation. The steps of the procedure of selection criteria for the evaluation are shown in Fig. 1.

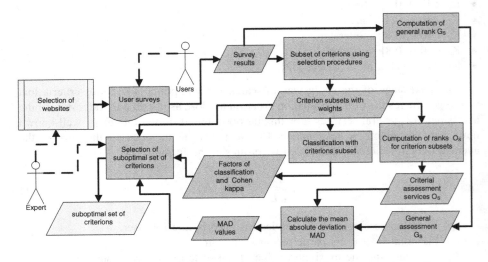

Fig. 1. The steps of the procedure of selection criteria for assessing the quality

4   Research Results

The first stage of the research was to gather user surveys for the assessment of the quality of the set of news portals. The survey included questions that correspond to each evaluation criteria using functioning models such as eQual, Ahn, SiteQual, WPSQ and WEQ based on top sites from national ranking[3]. In addition, the survey included a question regarding an overall rating of each service. For comparison purposes, users were also asked about their preferences explicitly declared as the weights of the criteria. Therefore, the users gave answers about how they evaluated each of the websites in terms of a total of 67 evaluation criteria; how important it was for them to have each of the criteria satisfied and how to generally evaluate each of the sites. The study used a seven-step Likert scale. Due to the very large number of questions, filling out the survey in one session would be very tedious and could result in unreliable responses to questions contained in it. Therefore, the survey was divided into three parts and accessed by the users at weekly intervals. The study collected 133 questionnaires containing criteria evaluations and overall assessments of each site; therefore, base 532 objects were available. In the next step feature selection procedures were applied,

---

[3] http://www.gemius.pl/pl/aktualnosci/2014-02-17/01.

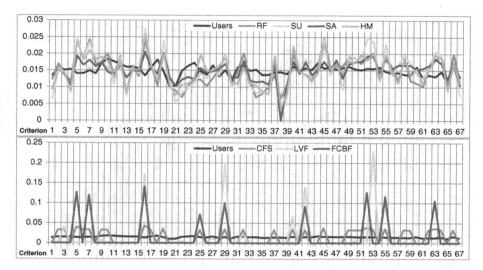

Fig. 2. Significance of particular criteria obtained using feature selection procedures (Color figure online)

making it possible to get rankings and relevance of the criteria. Normalized to the range [0, 1], significance criteria in a graphical form are shown in Fig. 2. Figure 2 separately shows the results obtained using procedures CFS LVF and FCBF to maintain its relative transparency. The complete results of this process are shown in Appendix 1 with columns C for criterion and S for significance. By using the FCBF, CFS and LVF methods, reduced sets of criteria and their weights were obtained, whereas the ReliefF, SU and SA methods yielded the rankings of criteria together with weights.

The analysis of Fig. 2 and Appendix 1 reveals several consistencies. A high level of similarity can be observed among all of the rankings based on Symmetrical Uncertainty, i.e. FCBF, CFS and SU. Additionally, all of the procedures for the selection of features for the least important criterion considered K38. Similarly, in all the rankings, among others, low-ranking features include: K21, K36, K34, K35 and K39. This succession coincides with reduced sets of criteria: FCBF, CFS and LVF, which do not include this subset of criteria. In turn, some of the most important criteria in each of the rankings are K16, K5, K19, K45 and K53. These results differ significantly from the ranking obtained on the basis of users' expressed responses in surveys, such as the features K16 and K63, which the users indicated were not very important compared to the rest of the features. Additionally, feature K13 which, according to the users, was considered one of the most important features, is positioned low in other rankings. There are also marked differences between the rankings created using various procedures for the selection of features. For example, the K52 feature was considered very important by the procedure Symmetrical Uncertainty, CFS, FCBF and Attribute Significance, while in the ReliefF, LVF and Hellwig rankings it occupies lower positions.

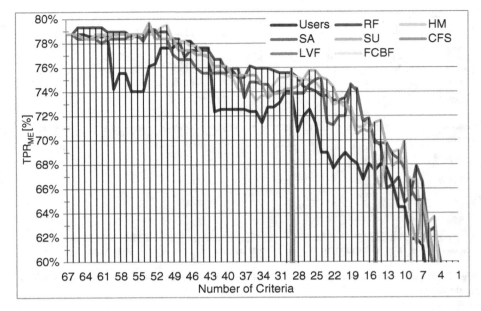

Fig. 3. Mean True Positive Rate for the particular subsets of criteria (Color figure online)

After creating rankings, the characteristics were tested using the decision tree CART. For each subset of criteria a true positive rate for each class of decision-making and decision-making for all classes together was set. The complete subsets designated by FCBF, LVF and CFS methods were also included. An average true positive rates for subsets of criteria obtained with the used selection procedures are showed in Fig. 3.

Moreover, for subsets based on the confusion matrix, Cohen's kappa value was determined (K_C). For the K_C the standard error was set and on this basis, the confidence interval was 0.99 [28, pp. 543–547]. The confusion matrix for the classification carried out using the full set of 67 criteria was characterized by a pointer value $K_C = 0.738$. Table 6 contains the value of the lower limit of the confidence interval of Cohen's kappa (K_{Cmin}) and the true positive rate for the worst classified class (TPR_{MIN}). Subsets of criteria satisfying the conditions $TRP_{MIN} > 50\%$ and $K_{Cmin} > 0.6$ are marked with a pattern.

The results presented in Table 6 and Fig. 3 show that the subsets of criteria created by the feature selection procedures allow for a more correct classification than the plurality of subsets of the same criteria, created based on the ranking of the users. Hence, it can be concluded that each of the procedures for the selection of features allows users to better specify the criteria that determine the quality of websites.

The next stage of the study consisted in assigning weights to criteria and calculating the mean absolute deviation according to Eq. (3). The MAD value for the subset of all criteria is shown in Fig. 4. For MAD values for a subset of stances from 9 to 30, the

Table 6. K_{Cmin} and TPR_{MIN} for subsets of criteria

Features	Users K_{Cmin}	Users TPR MIN [%]	ReliefF K_{Cmin}	ReliefF TPR MIN [%]	Symmetric Uncertainty K_{Cmin}	Symmetric Uncertainty TPR MIN [%]	Significance Attribute K_{Cmin}	Significance Attribute TPR MIN [%]	Hellwig Method K_{Cmin}	Hellwig Method TPR MIN [%]	CFS K_{Cmin}	CFS TPR MIN [%]	LVF K_{Cmin}	LVF TPR MIN [%]	FCBF K_{Cmin}	FCBF TPR MIN [%]
9	0.458	22.86	0.507	37.04	0.483	31.43	0.518	8.57	0.527	45.71					0.488	31.43
10	0.495	34.29	0.501	37.04	0.567	40.00	0.537	25.71	0.535	45.71						
11	0.496	34.29	0.529	51.43	0.545	40.00	0.547	25.71	0.558	45.71						
12	0.516	28.57	0.522	51.43	0.536	0.00	0.553	34.29	0.557	51.43						
13	0.538	47.06	0.519	59.26	0.558	25.71	0.564	44.44	0.561	44.44						
14	0.542	43.53	0.565	57.14	0.591	42.86	0.564	44.44	0.514	17.14						
15	0.537	51.85	0.570	57.14	0.589	42.86	0.563	37.14	0.528	17.14			0.558	25.71		
16	0.542	45.71	0.590	60.00	0.578	42.86	0.589	37.14	0.589	37.14						
17	0.528	62.96	0.592	60.00	0.581	42.86	0.588	51.43	0.582	37.14						
18	0.545	57.14	0.625	60.00	0.575	42.86	0.620	37.14	0.582	37.14						
19	0.551	62.96	0.626	60.00	0.595	51.43	0.629	54.29	0.603	45.71						
20	0.557	62.96	0.613	54.29	0.608	54.29	0.596	57.14	0.603	45.71						
21	0.550	48.15	0.610	54.29	0.608	54.29	0.596	57.14	0.608	51.43						
22	0.540	48.15	0.611	54.29	0.625	54.29	0.586	62.86	0.607	51.43						
23	0.558	60.00	0.615	54.29	0.633	54.29	0.589	62.86	0.622	54.29						
24	0.558	60.00	0.616	57.14	0.636	54.29	0.636	62.86	0.624	54.29						
25	0.585	51.43	0.620	57.14	0.643	57.14	0.634	62.86	0.632	57.14						
26	0.599	51.43	0.623	57.14	0.643	57.14	0.629	62.86	0.633	51.43						
27	0.592	51.43	0.626	57.14	0.634	62.86	0.619	60.00	0.633	51.43						
28	0.574	40.00	0.635	57.14	0.626	56.47	0.619	60.00	0.635	51.43						
29	0.620	62.86	0.640	57.14	0.623	56.47	0.619	60.00	0.637	51.43	0.645	62.86				
30	0.620	62.86	0.640	57.14	0.623	56.47	0.619	60.00	0.634	62.86						
67	0.682	60.00	0.682	60.00	0.682	60.00	0.682	60.00	0.682	60.00						

criteria are shown in Table 7. The results show retained TRP_{MIN} subsets which satisfy the conditions of > 50 % and K_{Cmin} > 0.6.

It should be noted that usually the mean absolute deviation obtained for a subset of the criteria established on the basis of users is much higher than for other subsets. Moreover, for all subsets created using feature selection procedures which meet the

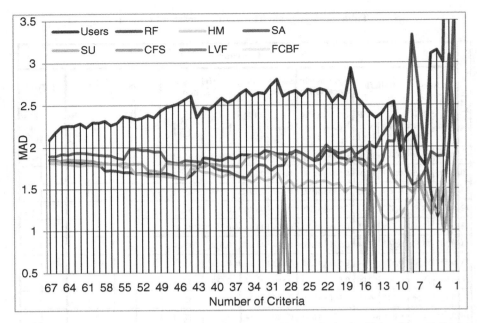

Fig. 4. Mean Absolute Deviation for the particular subsets of criteria (Color figure online)

conditions $TRP_{MIN} > 50\%$ and $K_{Cmin} > 0.6$, the value of MAD is lower than the value obtained for the full set of features with weights which are the average weights given by users (users' 67 criteria).

5   Discussion

According to the linguistic interpretation, there was substantial agreement between the observed and predicted values of the classifier. After taking into account the confidence level 0.99 and the calculation of the standard error for the indicator $K_C = 0.738$, the confidence interval [0.682, 0.794] was obtained. After examining the ratio K_C for the full set of criteria, it is assumed that the reduction criteria for the lower limit of the confidence interval (determined in the same manner as for the full set of criteria) should reach a value higher than 0.6. Based on the interpretation of the linguistic values of K_C, this means that consistency between the observed and predicted values of the classifier, with a probability of 99%, is maintained at a substantial level. In addition, it is assumed that the true positive rate for the worst classified class (TPR_{MIN}) should reach a value of over 50%. For a subset of the criteria obtained through the use of the feature selection method, these values were achievable for a subset of up to about 20 criteria. In turn, for subsets based on the weights assigned by the user criteria, the parameters of such subsets number approximately 30 criteria. Table 7 shows sets of criteria ranging

Table 7. Values of Mean Absolute Deviation for criteria subsets

Features	Users	ReliefF	Symmetric uncertainty	Significance attribute	Hellwig method	CFS	LVF	FCBF
9	2.12	2.30	1.51	1.70	1.28			1.37
10	1.94	2.32	1.51	2.36	1.17			
11	2.54	2.37	1.60	2.06	1.14			
12	2.50	2.23	1.79	2.07	1.11			
13	2.40	2.13	1.74	1.83	1.18			
14	2.34	1.99	1.76	1.71	1.39			
15	2.41	2.02	1.81	1.73	1.50		2.05	
16	2.51	1.97	1.77	1.84	1.47			
17	2.59	1.92	1.83	1.86	1.50			
18	2.94	1.82	1.86	1.98	1.52			
19	2.57	1.86	1.78	1.94	1.45			
20	2.62	1.86	1.81	1.92	1.56			
21	2.53	1.94	1.79	1.96	1.54			
22	2.67	1.96	1.79	2.02	1.59			
23	2.69	1.85	1.71	1.92	1.59			
24	2.66	1.83	1.77	1.84	1.57			
25	2.68	1.88	1.77	1.87	1.60			
26	2.61	1.92	1.81	1.92	1.54			
27	2.67	1.95	1.87	1.95	1.50			
28	2.65	1.90	1.89	1.93	1.60			
29	2.60	1.91	1.84	1.78	1.54	1.52		
30	2.80	1.91	1.89	1.77	1.68			
67	2.09	1.85	1.85	1.90	1.81			

from 9 to 30 criteria. This analysis confirms the theory which states that, depending on which feature selection procedure is used, better identification of the relevant quality criteria and their weights can be found than from the users themselves. Based on the value of MAD, TPR_{MIN} and K_{Cmin} selected subset of criteria characterizing the corresponding values of these parameters and the small cardinality criteria. The selected subset of a 21-piece set of criteria was created using individual indicators' information capacity. This set complies with the terms of the coefficients TPR_{MIN} and K_{Cmin}, and is further characterized by a small number of the criteria used and one of the lowest values of the received MAD. Among other subsets that meet the established conditions, a slightly lower value of MAD is characterized by only a 27-piece set of criteria created using the Hellwig method. A much lower value of the mean absolute deviation, however, does not compensate for a greater multiplicity of evaluation criteria.

6   Summary

The presented research explores the issues concerned with website quality assessment, reduction of assessment criteria and the determination of their weights using the feature selection methods. In today's heavily information-reliant society, in which most official matters and commercial transactions are conducted through the Internet, the quality and usability of a website is highly significant to the users' perception of an organization. Given the number of websites available online, their designers deem it important to maintain the number of Internet users visiting their sites, and this can be achieved only by ensuring the high quality of websites. This article aims to formulate a procedure enabling a reduction of the number of website assessment criteria without affecting the outcome of such assessment. The study demonstrated the results of research on the applicability of the feature selection and data mining methods in the developed procedure. According to the study, the developed procedure produces more reliable results from the multi-criteria assessment than those obtained using the full set of criteria. This is achieved through the procedure, which defines the implicit preferences of users in determining a website's quality. Although the indicated procedure was illustrated using one example of a website quality assessment model, it can be successfully transferred onto any type of model for assessing websites.

The proposed procedure for selection criteria, which uses machine learning methods, allowed the prediction of a suboptimal subset of criteria for assessing the quality of information services. This subset of the full set of criteria and their weights specified by the users allowed for a more precise multicriteria assessment of the information services. During the selection criteria it was also shown that declarations of the users' opinions regarding how important the criteria were did not correspond to the actual weights of the criteria. In other words, users were subconsciously guided by criteria other than those explicitly declared in their evaluation of the quality of websites. It can therefore be concluded that the developed procedure determines the implicit preferences that guide the users, assessing the true quality of the Internet service. Although for the indicated method of procedure only information services were presented as an example, the method can be adjusted to the designated criteria and their weights for the evaluation of other types of websites.

Future work should include additional analysis related to using internal system parameters based on automated measurements. The next stages should include the implementation of the developed procedure together with ontologies describing qualitative models and methods for a Multi Criteria Decision Support System targeted to assessing the quality of websites.

Appendix 1

Rankings criteria obtained using feature selection procedures

Features	Procedure															
	Users		ReliefF		Symmetrical Uncertainty		Significance Attribute		Hellwig Method		CFS		LVF		FCBF	
	C	S	C	S	C	S	C	S	C	S	C	S	C	S	C	S
1	K10	5.804511	K45	0.0797	K16	0.1912	K16	0.541	K16	0.0259	K16	0.809	K53	0.628	K16	0.1912
2	K18	5.75188	K7	0.078	K52	0.173	K19	0.533	K5	0.0238	K5	0.754	K29	0.547	K52	0.173
3	K11	5.578947	K16	0.0771	K5	0.1727	K52	0.511	K45	0.023	K52	0.719	K16	0.468	K5	0.1727
4	K9	5.496241	K5	0.0759	K45	0.1696	K5	0.509	K40	0.0222	K45	0.698	K42	0.383	K7	0.1627
5	K44	5.466165	K53	0.0653	K19	0.166	K53	0.503	K7	0.0210	K63	0.68	K55	0.278	K55	0.159
6	K24	5.458647	K32	0.0647	K53	0.1628	K55	0.488	K53	0.0202	K19	0.667	K40	0.18	K63	0.1443
7	K12	5.330827	K19	0.0644	K7	0.1627	K63	0.485	K19	0.0199	K55	0.657	K3	0.109	K29	0.1365
8	K66	5.300752	K10	0.0641	K55	0.159	K47	0.476	K25	0.0198	K40	0.649	K32	0.062	K42	0.1255
9	K27	5.225564	K66	0.064	K40	0.1562	K7	0.475	K55	0.0197	K29	0.642	K27	0.036	K25	0.0979
10	K32	5.180451	K40	0.0639	K63	0.1443	K32	0.47	K51	0.0197	K53	0.637	K9	0.021		
11	K23	5.165414	K29	0.0634	K58	0.1369	K51	0.466	K63	0.0196	K7	0.632	K50	0.013		
12	K40	5.135338	K55	0.063	K29	0.1365	K40	0.465	K37	0.0193	K64	0.629	K22	0.009		
13	K49	5.120301	K12	0.0619	K49	0.1359	K45	0.464	K9	0.0189	K32	0.625	K58	0.006		
14	K60	5.082707	K64	0.0615	K64	0.1338	K50	0.461	K29	0.0187	K66	0.622	K7	0.004		
15	K14	5.067669	K52	0.0612	K12	0.1306	K42	0.46	K58	0.0186	K3	0.619	K14	0.002		
16	K13	5.06015	K50	0.0605	K50	0.129	K9	0.456	K64	0.0185	K58	0.616				
17	K31	5.06015	K49	0.0603	K32	0.129	K64	0.45	K8	0.0183	K42	0.613				
18	K54	5.045113	K6	0.06	K51	0.1283	K29	0.447	K52	0.0181	K17	0.612				
19	K17	5.037594	K58	0.0592	K6	0.1276	K49	0.445	K6	0.0181	K49	0.61				
20	K46	5.022556	K63	0.059	K47	0.1269	K66	0.442	K10	0.0178	K6	0.61				
21	K48	4.977444	K9	0.0563	K8	0.1269	K58	0.437	K49	0.0178	K37	0.609				
22	K42	4.969925	K51	0.0547	K42	0.1255	K57	0.437	K12	0.0177	K10	0.608				
23	K2	4.954887	K2	0.0544	K37	0.1231	K37	0.437	K66	0.0173	K2	0.607				
24	K52	4.954887	K8	0.0537	K9	0.1194	K6	0.436	K50	0.0172	K51	0.606				
25	K45	4.93985	K48	0.0523	K66	0.1179	K12	0.435	K47	0.0170	K59	0.606				
26	K53	4.917293	K14	0.0522	K59	0.117	K59	0.435	K28	0.0164	K25	0.605				
27	K15	4.909774	K44	0.0521	K57	0.1149	K62	0.435	K17	0.0163	K50	0.605				
28	K50	4.887218	K17	0.0521	K62	0.1148	K8	0.424	K32	0.0159	K62	0.605				
29	K47	4.842105	K47	0.0521	K17	0.1131	K25	0.411	K60	0.0155	K9	0.605				
30	K4	4.834586	K46	0.0514	K10	0.1098	K61	0.411	K62	0.0154						
31	K34	4.819549	K56	0.0512	K14	0.1085	K28	0.408	K46	0.0153						
32	K43	4.81203	K62	0.05	K2	0.1085	K17	0.408	K15	0.0152						
33	K51	4.789474	K28	0.0492	K27	0.1072	K56	0.406	K14	0.0152						
34	K29	4.781955	K31	0.048	K44	0.102	K60	0.402	K2	0.0150						
35	K3	4.759398	K41	0.0469	K3	0.0999	K2	0.396	K27	0.0150						
36	K41	4.729323	K42	0.0441	K25	0.0979	K46	0.396	K56	0.0147						
37	K7	4.714286	K37	0.0437	K41	0.0977	K44	0.394	K26	0.0147						
38	K33	4.714286	K11	0.0428	K60	0.0969	K14	0.394	K41	0.0146						
39	K22	4.699248	K3	0.0427	K15	0.0966	K15	0.388	K3	0.0141						
40	K25	4.661654	K15	0.0422	K56	0.0958	K27	0.387	K44	0.0141						
41	K55	4.639098	K27	0.0408	K46	0.0948	K10	0.381	K42	0.0139						
42	K64	4.639098	K43	0.04	K28	0.0936	K26	0.378	K42	0.0137						
43	K19	4.616541	K24	0.0395	K20	0.0897	K54	0.376	K59	0.0134						
44	K38	4.586466	K25	0.0393	K26	0.0876	K41	0.371	K11	0.0127						
45	K37	4.571429	K1	0.0393	K61	0.087	K18	0.367	K24	0.0125						
46	K57	4.541353	K33	0.0387	K18	0.0828	K3	0.367	K54	0.0123						
47	K56	4.488722	K57	0.0385	K54	0.0812	K43	0.362	K18	0.0120						
48	K6	4.458647	K59	0.0384	K43	0.0811	K11	0.359	K43	0.0119						
49	K5	4.443609	K54	0.0384	K24	0.077	K23	0.342	K33	0.0113						
50	K8	4.43609	K4	0.0382	K11	0.0756	K31	0.34	K31	0.0112						
51	K39	4.421053	K65	0.0371	K33	0.0706	K13	0.336	K30	0.0112						
52	K58	4.37594	K67	0.0365	K23	0.0694	K20	0.335	K57	0.0112						
53	K26	4.345865	K60	0.0359	K31	0.0692	K48	0.334	K67	0.0111						
54	K36	4.330827	K39	0.0346	K30	0.0692	K30	0.333	K23	0.011						
55	K35	4.315789	K30	0.0339	K48	0.0644	K24	0.326	K61	0.0108						
56	K62	4.315789	K26	0.0332	K4	0.0613	K65	0.324	K22	0.0107						
57	K59	4.300752	K18	0.0331	K35	0.0598	K22	0.321	K34	0.0105						
58	K1	4.270677	K61	0.0329	K63	0.0593	K33	0.316	K20	0.0105						
59	K16	4.24812	K23	0.0329	K34	0.0578	K35	0.309	K1	0.0095						
60	K63	4.172932	K20	0.0328	K22	0.057	K34	0.303	K65	0.0095						
61	K28	4.165414	K36	0.032	K36	0.0542	K39	0.289	K35	0.0088						
62	K67	4.120301	K21	0.0284	K39	0.054	K36	0.286	K4	0.0085						
63	K61	4	K13	0.0267	K67	0.0525	K67	0.278	K39	0.0083						
64	K30	3.796992	K35	0.0258	K13	0.051	K4	0.278	K13	0.0082						
65	K65	3.511278	K34	0.0224	K1	0.0443	K21	0.269	K36	0.008						
66	K21	3.278195	K22	0.0224	K21	0.0381	K1	0.248	K21	0.0075						
67	K20	3.218045	K38	0.0215	K38	0	K38	0	K38	0.0037						

References

1. Kim, S., Stoel, L.: Dimensional hierarchy of retail website quality. Inf. Manag. **41**, 619–633 (2004)
2. Jankowski, J.: Analysis of multiplayer platform users activity based on the virtual and real time dimension. In: Datta, A., Shulman, S., Zheng, B., Lin, S.-D., Sun, A., Lim, E.-P. (eds.) SocInfo 2011. LNCS, vol. 6984, pp. 312–315. Springer, Heidelberg (2011)
3. Chiou, W.C., Lin, C.C., Perng, C.: A strategic framework for website evaluation based on a review of the literature from 1995–2006. Inf. Manag. **47**, 282–290 (2010)
4. Grigoroudis, E., Litos, C., Moustakis, V.A., Politis, Y., Tsironis, L.: The assessment of user-perceived web quality: application of a satisfaction benchmarking approach. Eur. J. Oper. Res. **187**, 1346–1357 (2008)
5. Barnes, S.J., Vidgen, R.: The eQual approach to the assessment of e-commerce quality: a longitudinal study of internet bookstories. In: Suh, W. (ed.) Web Engineering: Principles and Techniques, pp. 161–181. Idea Group Publishing, Hershey (2005)
6. Ahn, T., Ryu, S., Han, I.: The impact of Web quality and playfulness on user acceptance of online retailing. Inf. Manag. **44**, 263–275 (2007)
7. Webb, H.W., Webb, L.A.: SiteQual: an integrated measure of Web site quality. J. Enterp. Inf. Manag. **17**, 430–440 (2004)
8. Yang, Z., Cai, S., Zhou, Z., Zhou, N.: Development and validation of an instrument to measure user perceived service quality of information presenting Web Portals. Inf. Manag. **42**, 575–589 (2005)
9. Elling, S., Lentz, L., de Jong, M., van den Bergh, H.: Measuring the quality of governmental websites in a controlled versus an online setting with the 'Website Evaluation Questionnaire'. Gov. Inf. Quart. **29**, 383–393 (2012)
10. Holzinger, A.: Usability engineering methods for software developers. Commun. ACM **48**, 71–74 (2005)
11. Jankowski, J.: Integration of collective knowledge in Fuzzy models supporting Web design process. In: Jędrzejowicz, P., Nguyen, N.T., Hoang, K. (eds.) ICCCI 2011, Part II. LNCS, vol. 6923, pp. 395–404. Springer, Heidelberg (2011)
12. Ziemba, P., Piwowarski, M., Jankowski, J., Wątróbski, J.: Method of criteria selection and weights calculation in the process of Web projects evaluation. In: Hwang, D., Jung, J.J., Nguyen, N.-T. (eds.) ICCCI 2014. LNCS, vol. 8733, pp. 684–693. Springer, Heidelberg (2014)
13. Chou, W.C., Cheng, Y.: A hybrid Fuzzy MCDM approach for evaluating website quality of professional accounting firms. Expert Syst. Appl. **39**, 2783–2793 (2012)
14. ISO/IEC 25010:2010(E): Systems and software engineering — Systems and software Quality Requirements and Evaluation (SQuaRE) — System and software quality models
15. Sorum, H., Andersen, K.N., Clemmensen, T.: Website quality in government: exploring the webmaster's perception and explanation of website quality. Transforming Gov. People Process Policy **7**, 322–341 (2013)
16. Kaya, T.: Multi-attribute evaluation of website quality in e-business using an integrated Fuzzy AHPTOPSIS methodology. Int. J. Comput. Intell. Syst. **3**, 301–314 (2010)
17. Albert, B., Tullis, T., Tedesco, D.: Beyond The Usability Lab, Conducting Large-Scale Online User Experience Studies. Morgan Kaufmann, Burlington (2010)
18. Rubin, J., Chisnell, D.: Handbook of Usability Testing, How to Plan, Design, and Conduct Effective Tests, 2nd edn. Wiley, Indianapolis (2008)
19. Nielsen, J.: Usability Engineering. Morgan Kaufmann, San Francisco (1993)

20. Nielsen, J.: Usability 101: Introduction to Usability. Jakob Nielsen's Alertbox, 4 January 2012. http://www.nngroup.com/articles/usability-101-introduction-to-usability/
21. ISO 9126-1:2001(E): Software engineering – Product quality – Part 1: Quality model
22. Hasan, L., Abuelrub, E.: Assessing the quality of web sites. Appl. Comput. Inform. **9**, 11–29 (2011)
23. Yang, Z., Cai, S., Zhou, Z., Zhou, N.: Development and validation of an instrument to measure user perceived service quality of information presenting Web Portals. Inf. Manag. **42**, 575–589 (2005)
24. Chmielarz, W.: Quality assessment of selected bookselling websites. Pol. J. Manag. Stud. **1**, 127–146 (2010)
25. Lin, H.F.: An application of fuzzy AHP for evaluating course website quality. Comput. Educ. **54**, 877–888 (2010)
26. Ho, C., Lee, Y.: The development of an e-travel service quality scale. Tour. Manag. **28**, 1434–1449 (2007)
27. Ou, C.X., Sia, C.L.: Consumer trust and distrust: an issue of website design. Int. J. Hum. Comput. Stud. **68**, 913–934 (2010)
28. Hwang, J., Yoon, Y.S., Park, N.H.: Structural effects of cognitive and affective responses to web advertisements, website and brand attitudes, and purchase intentions: the case of casual-dining restaurants. Int. J. Hospitality Manag. **30**, 897–907 (2011)
29. Yang, Q., Shao, J., Scholz, M., Plant, C.: Feature selection methods for characterizing and classifying adaptive Sustainable Flood Retention Basins. Water Res. **45**, 993–1004 (2011)
30. Zenebe, A., Zhou, L., Norcio, A.F.: User preferences discovery using Fuzzy models. Fuzzy Sets Syst. **161**, 3044–3063 (2010)
31. Ziemba, P., Piwowarski, M.: Procedure for selecting significant website quality evaluation criteria based on feature selection methods. Stud. Proc. Pol. Assoc. Knowl. Manag. **67**, 119–133 (2013)
32. Ziemba, P., Piwowarski, M.: Procedure of reducing website assessment criteria and user preference analyses. Found. Comput. Decis. Sci. **36**(3–4), 315–325 (2011)
33. Chizi, B., Maimon, O.: Dimension reduction and feature selection. In: Maimon, O., Rokach, L. (eds.) Data Mining and Knowledge Discovery Handbook, pp. 83–100. Springer, New York (2010)
34. Guyon, I.: Practical feature selection: from correlation to causality. In: Fogelman-Soulié, F., Perrotta, D., Piskorski, J., Steinberger, R. (eds.) Mining massive data sets for security: advances in data mining, search, social networks and text mining, and their applications to security, pp. 27–43. IOS Press, Amsterdam (2008)
35. Hand, D., Mannila, H., Smyth, D.: Eksploracja danych, pp. 414–416. WNT, Warszawa (2005)
36. Witten, I.H., Frank, E.: Data Mining. Practical Machine Learning Tools and Techniques, pp. 288–295. Morgan Kaufmann, San Francisco (2005)
37. Hall, M.A., Holmes, G.: Benchmarking attribute selection techniques for discrete class data mining. IEEE Trans. Knowl. Data Eng. **15**, 1437–1447 (2003)
38. Fu, H., Xiao, Z., Dellandréa, E., Dou, W., Chen, L.: Image categorization using ESFS: a new embedded feature selection method based on SFS. In: Blanc-Talon, J., Philips, W., Popescu, D., Scheunders, P. (eds.) ACIVS 2009. LNCS, vol. 5807, pp. 288–299. Springer, Heidelberg (2009)
39. Hsu, H.H., Hsieh, C.W., Lu, M.D.: Hybrid feature selection by combining filters and wrappers. Expert Syst. Appl. **38**, 8144–8150 (2011)
40. Chang, C.C.: Generalized iterative RELIEF for supervised distance metric learning. Pattern Recogn. **43**, 2971–2981 (2010)

41. Kononenko, I., Hong, S.J.: Attribute selection for modelling. Future Gener. Comput. Syst. **13**, 181–195 (1997)
42. Liu, H., Yu, L., Motoda, H.: Feature extraction, selection, and construction. In: Ye, N. (ed.) The Handbook of Data Mining, pp. 409–424. Lawrence Erlbaum Associates, Mahwah (2003)
43. Ahmad, A., Dey, L.: A feature selection technique for classificatory analysis. Pattern Recogn. Lett. **26**, 43–56 (2005)
44. Yu, L., Liu, H.: Feature selection for high-dimensional data: a fast correlation-based filter solution. In: Proceedings of the 20th International Conference on Machine Leaning (ICML 2003), pp. 856–863 (2003)
45. Hall, M.A.: Correlation-based feature selection for discrete and numeric class machine learning. In: Proceedings of the 17th International Conference on Machine Learning (ICML 2000), pp. 359–366 (2000)
46. Hellwig, Z.: On the optimal choice of predictors. In: Gostkowski, Z. (ed.) Toward a System of Quantitative Indicators of Components of Human Resources Development, Study VI. UNESCO, Paris (1968)
47. Senthamarai Kannan, S., Ramaraj, N.: A novel hybrid feature selection via Symmetrical Uncertainty ranking based local memetic search algorithm. Knowl.-Based Syst. **23**, 580–585 (2010)
48. Rokach, L., Maimon, O.: Classification trees. In: Maimon, O., Rokach, L. (eds.) Data Mining and Knowledge Discovery Handbook, 2nd edn, pp. 149–174. Springer, New York (2010)
49. Webb, G.I.: Association rules. In: Ye, N. (ed.) The Handbook of Data Mining, pp. 25–40. Lawrence Erlbaum Associates, Mahwah (2003)
50. Rokach, L., Maimon, O.: Supervised learning. In: Maimon, O., Rokach, L. (eds.) Data Mining and Knowledge Discovery Handbook, 2nd edn, pp. 133–148. Springer, New York (2010)
51. Ben-David, A.: Comparison of classification accuracy using Cohen's Weighted Kappa. Expert Syst. Appl. **34**, 825–832 (2008)
52. Kuchenhoff, H., Augustin, T., Kunz, A.: Partially identified prevalence estimation under misclassification using the kappa coefficient. Int. J. Approximate Reasoning **53**, 1168–1182 (2012)
53. Pham-Gia, T., Hung, T.L.: The mean and median absolute deviations. Math. Comput. Model. **34**, 921–936 (2001)

Dynamic Database by Inconsistency and Morphogenetic Computing

Xiaolin Xu[1], Germano Resconi[2(✉)], and Guanglin Xu[3]

[1] College of International Vocational Education,
Shanghai Second Polytechnic University,
Shanghai, China
xlxu2001@163.com
[2] Department of Mathematics and Physics,
Catholic University, Brescia, Italy
resconi@speedyposta.it
[3] College of Mathematics and Information,
Shanghai Lixin University of Commerce,
Shanghai, China
glxu@outlook.com

Abstract. Since Peter Chen published the article *Entity–Relationship Modeling* in 1976, Entity-Relationship database has become a hot spot for research. With the advent of the big data, it appears that Entity-Relationship database is substituted for attribute reduction map structure. In the big data we have no evidence of the relationship but only of attributes and maps. In this paper we give an attribute representation of the relationship. In fact we assume that any entity can be in two different attributes (states) with two different values. One is the attribute that sends a message that we denote as e_1 and the other is to receive the message that we denote as e_2. The values of the attributes are the names of the entities. A relationship is a superposition $ae_1 + be_2$ of the two states. When we change the values of the states we change the database. When we change the two states in the same way we have isomorphism among database, and when we change the two states in different way we have isomorphism with distortion (homotopic transformation). Given a set of independent data base we can generate (compute) all the other data base in a dynamical way. In this way we can reduce the database that we must memorize. Because we are interested in the generation of the form (morphology) of database we denote this new model of computation as morphogenetic computing.

Keywords: Entity–Relationship model · Morphogenetic computing · Dynamical process · Inconsistency · Compensation

1 Introduction

Since Peter Chen published the article Entity–Relationship Modeling in 1976 [1], relational database appears to be popular in the research of database [2]. Some scholars carried out the studies from the point of Algebra [3]. Some scholars applied fuzzy set to relational database [4–7] and had developed relevant software products. Some scholars

© Springer-Verlag Berlin Heidelberg 2016
N.T. Nguyen and R. Kowalczyk (Eds.): TCCI XXII, LNCS 9655, pp. 189–204, 2016.
DOI: 10.1007/978-3-662-49619-0_10

focused on data warehouse and data mining in terms of multidimensional models [8–10] conducted a preliminary study of conflict compensation, redundancy and similarity from the perspective of morphogenetic computing which provided the possibility of the integration of databases to eliminate the redundancy. On the basis of [10], the main point of this paper is to present different dynamic types in database including the internal dynamic type and the meta dynamic type. The former is the case where information moves from one entity to another in a complex way that we can control by from/to table and by the association with graph database structure in a multidimensional space where we can pick up general property of the database structure as source, sink, transition and so on. With the internal property of the database it is possible to control all the information flux in a global way. The latter is a meta dynamic type where one database can be transformed into to another database with the same property or partially common property and one database can be taken as reference to create another database. In this way, the new database is not similar to the old one completely but has some improvement or deformation on the properties of the previous database. Thus in database design, we do not need to change completely the previous database but to generate suitable database with new necessities in agreement with the previous database. This adaptation cannot generate without conflicts and solutions to conflicts. We think that formal and mathematical description of a well known process that a lot of people use in an empirical way can give us a conceptual guide to obtain the adaptation to the new situation that changes in any time and any space in a more efficient and rapid way.

2 Representation of Relationship by Morphogenetic Computing

Given that $\{Entity_1, Entity_2, \ldots, Entity_n\}$ is set of entities, (1) represents the relationship matrix R between entities.

$$\begin{bmatrix} R & Entity_1 & Entity_2 & Entity_3 & \cdots & Entity_n \\ Entity_1 & e_{1,1} & e_{1,2} & e_{1,3} & \cdots & e_{1,n} \\ Entity_2 & e_{2,1} & e_{2,2} & e_{2,3} & \cdots & e_{2,n} \\ Entity_3 & e_{3,1} & e_{3,2} & e_{3,3} & \cdots & e_{3,n} \\ \cdots & \cdots & \cdots & \cdots & \cdots & \cdots \\ Entity_n & e_{n,1} & e_{n,2} & e_{n,3} & \cdots & e_{n,n} \end{bmatrix} \quad (1)$$

Where, $e_{i,j}$ is the relationship between $Entity_i$ and $Entity_j$. If there is a relation between $Entity_i$ and $Entity_j$, $e_{i,j}$ is 1, otherwise, $e_{i,j}$ is 0.

Example 1. There is an entity-relationship diagram including five entities showed in Fig. 1.

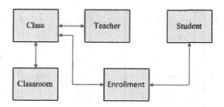

Fig. 1. Entity-relationship diagram with five entities

(2) shows the relationship matrix R for the relationships in Fig. 1.

$$
\begin{bmatrix}
R & class & classroom & enrollment & teacher & student \\
class & 0 & 1 & 1 & 1 & 0 \\
classroom & 1 & 0 & 0 & 0 & 0 \\
enrollment & 1 & 0 & 0 & 0 & 1 \\
teacher & 1 & 0 & 0 & 0 & 0 \\
student & 0 & 0 & 1 & 0 & 0
\end{bmatrix} \tag{2}
$$

Then it can be represented as (3).

$$
Rv = \begin{bmatrix}
0 & 1 & 1 & 1 & 0 \\
1 & 0 & 0 & 0 & 0 \\
1 & 0 & 0 & 0 & 1 \\
1 & 0 & 0 & 0 & 0 \\
0 & 0 & 1 & 0 & 0
\end{bmatrix}
\begin{bmatrix}
class \\
classroom \\
enrollment \\
teacher \\
student
\end{bmatrix}
=
\begin{bmatrix}
(classroom, enrollment, teacher) \\
class \\
(class, student) \\
class \\
enrollment
\end{bmatrix}
\tag{3}
$$

The difference between (2) and (3) is that (2) is a static representation and (3) is a dynamic representation where both the initial set of entities and the final set of entities can be seen. It is noticed that the initial set includes individual names and the final set could include some subsets of entities. That means one entity as the initial entity could be associated with other several entities, and the transformation can be a one-to-many process with intrinsic uncertainty that there are possibly different final entities from one initial entity. The transformation (3) can be written in (4) or (5).

$$Rv = \begin{bmatrix} 1 & 0 & 0 & 0 & 0 \\ 0 & 1 & 0 & 0 & 0 \\ 0 & 0 & 1 & 0 & 0 \\ 0 & 0 & 0 & 1 & 0 \\ 0 & 0 & 0 & 0 & 1 \end{bmatrix} \begin{bmatrix} class \\ classroom \\ enrollment \\ teacher \\ student \end{bmatrix} e_1 + \begin{bmatrix} 0 & 1 & 1 & 1 & 0 \\ 1 & 0 & 0 & 0 & 0 \\ 1 & 0 & 0 & 0 & 1 \\ 1 & 0 & 0 & 0 & 0 \\ 0 & 0 & 1 & 0 & 0 \end{bmatrix} \begin{bmatrix} class \\ classroom \\ enrollment \\ teacher \\ student \end{bmatrix} e_2$$

$$= \left(\begin{bmatrix} 1 & 0 & 0 & 0 & 0 \\ 0 & 1 & 0 & 0 & 0 \\ 0 & 0 & 1 & 0 & 0 \\ 0 & 0 & 0 & 1 & 0 \\ 0 & 0 & 0 & 0 & 1 \end{bmatrix} e_1 + \begin{bmatrix} 0 & 1 & 1 & 1 & 0 \\ 1 & 0 & 0 & 0 & 0 \\ 1 & 0 & 0 & 0 & 1 \\ 1 & 0 & 0 & 0 & 0 \\ 0 & 0 & 1 & 0 & 0 \end{bmatrix} e_2 \right) \begin{bmatrix} class \\ classroom \\ enrollment \\ teacher \\ student \end{bmatrix}$$

$$= \begin{bmatrix} (class)e_1 + (classroom, enrollment, teacher)e_2 \\ (classroom)e_1 + (class)e_2 \\ (enrollment)e_1 + (class, student)e_2 \\ (teacher)e_1 + (class)e_2 \\ (student)e_1 + (enrollment)e_2 \end{bmatrix}$$

$$(4)$$

$$\left(\begin{bmatrix} 1 & 0 & 0 & 0 & 0 \\ 0 & 1 & 0 & 0 & 0 \\ 0 & 0 & 1 & 0 & 0 \\ 0 & 0 & 0 & 1 & 0 \\ 0 & 0 & 0 & 0 & 1 \end{bmatrix} e_1 + \begin{bmatrix} 0 & 1 & 1 & 0 & 0 \\ 1 & 0 & 0 & 0 & 0 \\ 1 & 0 & 0 & 0 & 1 \\ 1 & 0 & 0 & 0 & 0 \\ 0 & 0 & 1 & 0 & 0 \end{bmatrix} e_2 \right) \begin{bmatrix} class \\ classroom \\ enrollment \\ teacher \\ student \end{bmatrix}$$

$$= \begin{bmatrix} e_1 & e_2 & e_2 & 0 & 0 \\ e_2 & e_1 & 0 & 0 & 0 \\ e_2 & 0 & e_1 & 0 & e_2 \\ e_2 & 0 & 0 & e_1 & 0 \\ 0 & 0 & e_2 & 0 & e_1 \end{bmatrix} \begin{bmatrix} class \\ classroom \\ enrollment \\ teacher \\ student \end{bmatrix} = \begin{bmatrix} (class)e_1 + (classroom + enrollment)e_2 \\ (classroom)e_1 + (class)e_2 \\ (enrollment)e_1 + (class + student)e_2 \\ (teacher)e_1 + (class)e_2 \\ (student)e_1 + (enrollment)e_2 \end{bmatrix}$$

$$(5)$$

Where we have the state e_1 that sends a message given by the teacher example.

$$\boxed{teacher} \xrightarrow{\;\;\;\;} {}_{e_1}$$

And the other is the state e_2 that receives a message given by the class example.

$$\boxed{class} \xleftarrow{\;\;\;\;} {}_{e_2}$$

The superposition can be represented as (6).

$$(teacher)e_1 + (class)e_2 \tag{6}$$

And or as Fig. 2.

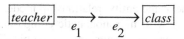

Fig. 2. The superposition from teacher to class

For the database we have all the representations of states given in Fig. 3.

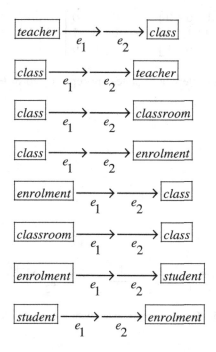

Fig. 3. All the superpositions in the database

We can superpose two relationships as Fig. 4. and represent the superposition as (7).

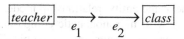

Fig. 4. The superposition of two relationships

$$(class)e_{11} + (enrolment)e_{21} + (enrolment)e_{12} + (student)e_{22} \tag{7}$$

3   Conflicts and Inconsistency of Relationship in Entity–Relationship Modeling and Compensation

The system axiom created by A.Wayne Wymore is that any system element or entity has one name [11], the same as the entity–relationship modeling created by Peter Chen [1]. But from the point of view of morphogenetic computing, any entity should have two or more conflicting names to change in a dynamical way in the form of the relationship (morphology) in given database. Thereby conflicts and inconsistency are likely to happen in relational database.

Example 2. Given the database with relationship as (8).

$$
\left(\begin{bmatrix} 1 & 0 & 0 \\ 0 & 1 & 0 \\ 0 & 0 & 1 \end{bmatrix} e_1 + \begin{bmatrix} 0 & 1 & 0 \\ 0 & 0 & 1 \\ 1 & 0 & 0 \end{bmatrix} e_2 \right) \begin{bmatrix} a \\ b \\ c \end{bmatrix}
\tag{8}
$$

The relationship is a cycle whose states are represented in Fig. 5.

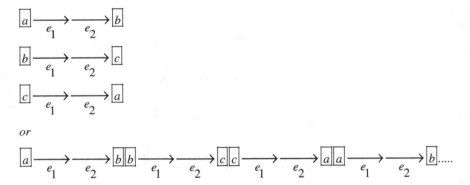

Fig. 5. The cycle represented with the superposition

Given the permutation of the values for the state e_1 we have the functor (Fig. 6).

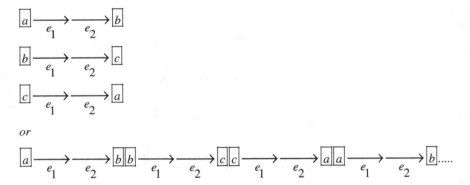

Fig. 6. The permutation of e_1

Now after the transformation if we do not want to change the relationship of the cycle, we have a conflicting situation because c goes to b now b changes name with a, a goes to c, c changes name and becomes b, b goes to a and again a changes name with c.

But the same entity has only one name so we have internal conflict to any entity. So we have the conflicting graph (Fig. 7).

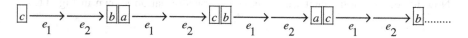

Fig. 7. The conflicting entity

When we transform the two states in the same way we have Fig. 8.

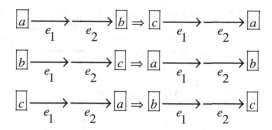

Fig. 8. The permutation of e_1 and e_2

So after the transformation we have the coherent graph Fig. 9.

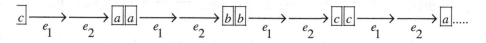

Fig. 9. The coherent entity

This means that we have again a coherent cycle without conflicting situation. In this case the transformation generates a new graph that has all properties of the previous cycle with the permuted entities. We remark that the two graphs (Figs. 5 and 9) are isomorphic and in any case we can only consider one graph and we can easily generate the second. In Fig. 5, the cycle begins with a. In Fig. 9, the cycle begins with c. However, the cycle is always the same.

For the inconsistent graph (Fig. 7) we can change the relationship in a way to establish a compensation (Fig. 10).

$$\boxed{c} \xrightarrow[e_1]{} \xrightarrow[e_2]{} \boxed{b}\boxed{b} \xrightarrow[e_1]{} \xrightarrow[e_2]{} \boxed{a}\boxed{a} \xrightarrow[e_1]{} \xrightarrow[e_2]{} \boxed{c}\boxed{c} \xrightarrow[e_1]{} \xrightarrow[e_2]{} \boxed{b}\dots$$

Fig. 10. The relationship after compensation

Now for the previous database we have the transformation shown in Fig. 11.

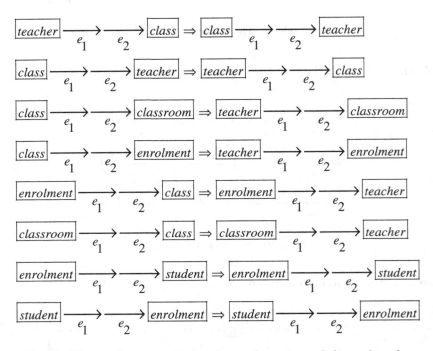

Fig. 11. The transformation with the change of two states of *class* and *teacher*

With the previous transformation the database changes the name *class* with the name *teacher* for the first state but also for the second state. So we can connect the entities with the same graph (Fig. 3) with the difference that we change only the name. In this case the new database is similar but not equal to the previous.

Now if we change *class* with *teacher* for only one state we have Fig. 12.

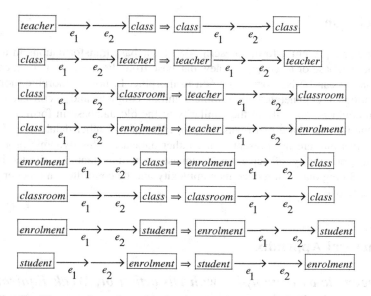

Fig. 12. The transformation with the change of one state of *class* and *teacher*

Thus the inconsistence occurs shown in Fig. 13.

$$\boxed{class} \xrightarrow[e_1]{} \xrightarrow[e_2]{} \boxed{class}\boxed{teacher} \xrightarrow[e_1]{} \xrightarrow[e_2]{} \boxed{teacher}$$

Fig. 13. The inconsistent superposition

With the transformation we create two new loops: *class* goes to *class* and *teacher* goes to *teacher*. *Teacher* and *class* now are disconnected. The new graph (Fig. 13) is not equal, not isomorphic to the original graph (Fig. 3), but we can make a distorsion (the two loops). After local distortion (Fig. 14), all the other parts of the database is equal to the previous.

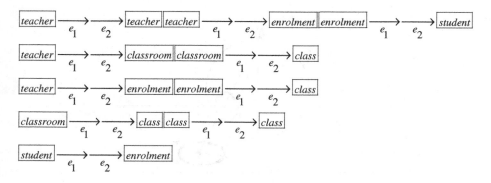

Fig. 14. The superpositions after distortion

4   Conclusion

In this paper we present a formal description of database transformations in a way to classify the database or to generate a new database from the previous known database. Transformation can be isomorphic or non-isomorphic. In non isomorphic, homotopic transformations are generated to eliminate initial conflicts and inconsistency. The generation could be not only in the similarity to the old database in form or structure but have the new database that owns the properties of the father database and adjoins new properties that are not present in the father database. This dynamic process also suggests the possibility of separating database in unities that are similar but also homotopic. So big data can reduce its complexity and be controlled in a better way by its homotopic parts.

Mathematical Appendix

Morphogenetic as Isomorphis with Distortion or, Weak Equivalence or Homotopy in Database

To understand the previous morphogenetic computing, we show different examples of the particular transformations denoted as isomorphism with deformation or homology and so on.

In the geographic database, there are two principal data of the earth geography. The first is located on a sphere and the other is the projection of the geography information on a plane. Now we know that the two databases are related but we also know that they are not isomorphic because when we project one image from the sphere, the new image on the plane has a distortion. In fact, in Map Projection Distorts Reality (topological similarity) a sphere is not a developable solid, and transfering from 3D globe to 2D map must result in the loss of one or global characteristics. Figure 15 shows the earth grid and features projected from sphere to a plane surface. Figure 16 shows the distorted map of the earth by planar projection.

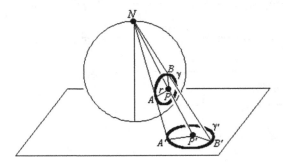

Fig. 15. The projection of a sphere on the plane

Fig. 16. Distorted map of the earth by planar projection

The distance is equidistance between sphere and plane projection. All the other elements are distorted. Given a path on the sphere, we project the initial point and the final point of the path into the plane. The path on the plane changes in a different way from one part of the sphere to another. The same path at the north pole has a little dimension but at the equator it is more or less the same on the sphere. We have no global transformation that changes paths from the sphere to the plane. We have only local transformation that changes point by point in the sphere. This is the gauge transformation due to the impossibility to have the curvature property of the sphere into the plane. When we move from the sphere to the plane, we lose properties because the plane has not curvature.

Formal Description of the Morphogenetic Computation

Morphogenetic computing is more than the abstract theory. In this part we make the formal morphogenetic computation on the dynamic relationships in database.

Given the relationship matrix R that reflects the from/to connection among entities.

$$R = \begin{bmatrix} r_{11} & r_{12} & r_{13} & \cdots & r_{1n} \\ r_{21} & r_{22} & r_{21} & \cdots & r_{2n} \\ r_{31} & r_{32} & r_{33} & \cdots & r_{3n} \\ \cdots & \cdots & \cdots & \cdots & \cdots \\ r_{n1} & r_{n2} & r_{n3} & \cdots & r_{nn} \end{bmatrix}$$

Then general operators A and B are created. Now the isomorphism with deformation is given by the formal expression (A.1).

$$\left[\begin{pmatrix} 1 & 0 & 0 & \cdots & 0 \\ 0 & 1 & 0 & \cdots & 0 \\ 0 & 0 & 1 & \cdots & 0 \\ \cdots & \cdots & \cdots & \cdots & \cdots \\ 0 & 0 & 0 & \cdots & 1 \end{pmatrix} \begin{pmatrix} a_{11} & a_{12} & a_{13} & \cdots & a_{1n} \\ a_{21} & a_{22} & a_{23} & \cdots & a_{2n} \\ a_{31} & a_{32} & a_{33} & \cdots & a_{3n} \\ \cdots & \cdots & \cdots & \cdots & \cdots \\ a_{n1} & a_{n2} & a_{n3} & \cdots & a_{nn} \end{pmatrix} e_1\right.$$

$$\left.+ \begin{pmatrix} b_{11} & b_{12} & b_{13} & \cdots & b_{1n} \\ b_{21} & b_{22} & b_{23} & \cdots & b_{2n} \\ b_{31} & b_{32} & b_{33} & \cdots & b_{3n} \\ \cdots & \cdots & \cdots & \cdots & \cdots \\ b_{n1} & b_{n2} & b_{n3} & \cdots & b_{nn} \end{pmatrix} \begin{pmatrix} r_{11} & r_{12} & r_{13} & \cdots & r_{1n} \\ r_{21} & r_{22} & r_{23} & \cdots & r_{2n} \\ r_{31} & r_{32} & r_{33} & \cdots & r_{3n} \\ \cdots & \cdots & \cdots & \cdots & \cdots \\ r_{n1} & r_{n2} & r_{n3} & \cdots & r_{nn} \end{pmatrix} e_2\right] \begin{bmatrix} name_1 \\ name_2 \\ name_3 \\ \cdots \\ name_3 \end{bmatrix} \quad\text{(A.1)}$$

$$= (Ae_1 + BRe_2)D$$

When A = B, we have (A.2).

$$(Ae_1 + ARe_2)D = A(e_1 + Re_2)D \tag{A.2}$$

In this case, the input graph and output graph are the same but names of the entities D change. So the morphogenetic result is an isomorphic database where the structure is the same but the names change.

When A ≠ B and B = A C we have (A.3).

$$(Ae_1 + BRe_2)D = (Ae_1 + ACRe_2)D = A(e_1 + CRe_2)D \tag{A.3}$$

This appears to be an isomorphism but with correction by C.

Example 1. For relationship R1 shown Fig. 17, we have the representation (A.4).

$$R_1 D = \left(\begin{pmatrix} 1 & 0 & 0 & 0 & 0 \\ 0 & 1 & 0 & 0 & 0 \\ 0 & 0 & 1 & 0 & 0 \\ 0 & 0 & 0 & 1 & 0 \\ 0 & 0 & 0 & 0 & 1 \end{pmatrix} e_1 + \begin{pmatrix} 0 & 1 & 1 & 1 & 0 \\ 1 & 0 & 0 & 0 & 0 \\ 1 & 0 & 0 & 0 & 1 \\ 1 & 0 & 0 & 0 & 0 \\ 0 & 0 & 1 & 0 & 0 \end{pmatrix} e_2\right) \begin{bmatrix} class \\ classroom \\ enrollment \\ teacher \\ student \end{bmatrix} \quad\text{(A.4)}$$

Given the permutations

$$A = \begin{pmatrix} 0 & 1 & 0 & 0 & 0 \\ 1 & 0 & 0 & 0 & 0 \\ 0 & 0 & 1 & 0 & 0 \\ 0 & 0 & 0 & 1 & 0 \\ 0 & 0 & 0 & 0 & 1 \end{pmatrix}, \; B = \begin{pmatrix} 1 & 0 & 0 & 0 & 0 \\ 0 & 1 & 0 & 0 & 0 \\ 0 & 0 & 1 & 0 & 0 \\ 0 & 0 & 0 & 0 & 1 \\ 0 & 0 & 0 & 1 & 0 \end{pmatrix}$$

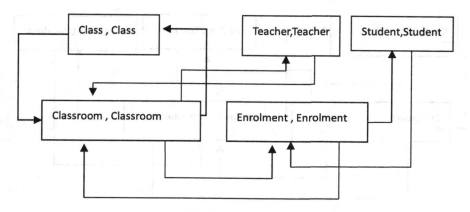

Fig. 17. Relation R_1

$$R_2 = Ae_1 + BR_1e_2 = (A \begin{bmatrix} 1 & 0 & 0 & 0 & 0 \\ 0 & 1 & 0 & 0 & 0 \\ 0 & 0 & 1 & 0 & 0 \\ 0 & 0 & 0 & 1 & 0 \\ 0 & 0 & 0 & 0 & 1 \end{bmatrix} e_1 + B \begin{bmatrix} 0 & 1 & 1 & 1 & 0 \\ 1 & 0 & 0 & 0 & 0 \\ 1 & 0 & 0 & 0 & 1 \\ 1 & 0 & 0 & 0 & 0 \\ 0 & 0 & 1 & 0 & 0 \end{bmatrix} e_2)$$

$$= (\begin{bmatrix} 0 & 1 & 0 & 0 & 0 \\ 1 & 0 & 0 & 0 & 0 \\ 0 & 0 & 1 & 0 & 0 \\ 0 & 0 & 0 & 1 & 0 \\ 0 & 0 & 0 & 0 & 1 \end{bmatrix} \begin{bmatrix} 1 & 0 & 0 & 0 & 0 \\ 0 & 1 & 0 & 0 & 0 \\ 0 & 0 & 1 & 0 & 0 \\ 0 & 0 & 0 & 1 & 0 \\ 0 & 0 & 0 & 0 & 1 \end{bmatrix} e_1 + \begin{bmatrix} 1 & 0 & 0 & 0 & 0 \\ 0 & 1 & 0 & 0 & 0 \\ 0 & 0 & 1 & 0 & 0 \\ 0 & 0 & 0 & 0 & 1 \\ 0 & 0 & 0 & 1 & 0 \end{bmatrix} \begin{bmatrix} 0 & 1 & 1 & 1 & 0 \\ 1 & 0 & 0 & 0 & 0 \\ 1 & 0 & 0 & 0 & 1 \\ 1 & 0 & 0 & 0 & 0 \\ 0 & 0 & 1 & 0 & 0 \end{bmatrix} e_2)$$

$$= (\begin{bmatrix} 0 & 1 & 0 & 0 & 0 \\ 1 & 0 & 0 & 0 & 0 \\ 0 & 0 & 1 & 0 & 0 \\ 0 & 0 & 0 & 1 & 0 \\ 0 & 0 & 0 & 0 & 1 \end{bmatrix} e_1 + \begin{bmatrix} 1 & 0 & 0 & 0 & 0 \\ 0 & 1 & 0 & 0 & 0 \\ 0 & 0 & 1 & 0 & 0 \\ 0 & 0 & 0 & 0 & 1 \\ 0 & 0 & 0 & 1 & 0 \end{bmatrix} \begin{bmatrix} 0 & 1 & 1 & 1 & 0 \\ 1 & 0 & 0 & 0 & 0 \\ 1 & 0 & 0 & 0 & 1 \\ 1 & 0 & 0 & 0 & 0 \\ 0 & 0 & 1 & 0 & 0 \end{bmatrix} e_2)$$

$$= \begin{bmatrix} 0 & 1 & 0 & 0 & 0 \\ 1 & 0 & 0 & 0 & 0 \\ 0 & 0 & 1 & 0 & 0 \\ 0 & 0 & 0 & 1 & 0 \\ 0 & 0 & 0 & 0 & 1 \end{bmatrix} e_1 + \begin{bmatrix} 0 & 1 & 1 & 1 & 0 \\ 1 & 0 & 0 & 0 & 0 \\ 1 & 0 & 0 & 0 & 1 \\ 0 & 0 & 1 & 0 & 0 \\ 1 & 0 & 0 & 0 & 0 \end{bmatrix} e_2$$

$$(A.5)$$

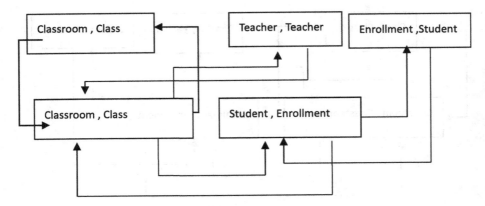

Fig. 18. The inconsistence after permutation

The inconsistent Fig. 18 shows the mixed names in one entity.
Particular case for the diagram

$$1)\ A = B = A\left(\begin{bmatrix} 1 & 0 & 0 & 0 & 0 \\ 0 & 1 & 0 & 0 & 0 \\ 0 & 0 & 1 & 0 & 0 \\ 0 & 0 & 0 & 1 & 0 \\ 0 & 0 & 0 & 0 & 1 \end{bmatrix} Ae_1 + A \begin{bmatrix} 0 & 1 & 1 & 1 & 0 \\ 1 & 0 & 0 & 0 & 0 \\ 1 & 0 & 0 & 0 & 1 \\ 1 & 0 & 0 & 0 & 0 \\ 0 & 0 & 1 & 0 & 0 \end{bmatrix} e_2\right)$$

$$= A\left(\begin{bmatrix} 1 & 0 & 0 & 0 & 0 \\ 0 & 1 & 0 & 0 & 0 \\ 0 & 0 & 1 & 0 & 0 \\ 0 & 0 & 0 & 1 & 0 \\ 0 & 0 & 0 & 0 & 1 \end{bmatrix} e_1 + \begin{bmatrix} 0 & 1 & 1 & 1 & 0 \\ 1 & 0 & 0 & 0 & 0 \\ 1 & 0 & 0 & 0 & 1 \\ 1 & 0 & 0 & 0 & 0 \\ 0 & 0 & 1 & 0 & 0 \end{bmatrix} e_2\right)$$

$$\text{For } A = \begin{bmatrix} 0 & 1 & 0 & 0 & 0 \\ 1 & 0 & 0 & 0 & 0 \\ 0 & 0 & 1 & 0 & 0 \\ 0 & 0 & 0 & 1 & 0 \\ 0 & 0 & 0 & 0 & 1 \end{bmatrix}$$

We have the coherent graph Fig. 19.
Multi-dimensional transformation is shown in Fig. 20.
The two dimensional reference space (e_{11}, e_{21}) can be expanded in four dimensional space reference (e_{11}, e_{12}, e_{21}, e_{22}). From the same entities D we have the first relationship structure, form or topology denoted R_{11}.

$$D = \begin{bmatrix} class \\ classroom \\ enrolment \\ teacher \\ student \end{bmatrix} \quad \text{and } R_{11} = \begin{bmatrix} 0 & 1 & 1 & 1 & 0 \\ 1 & 0 & 0 & 0 & 0 \\ 1 & 0 & 0 & 0 & 1 \\ 1 & 0 & 0 & 0 & 0 \\ 0 & 0 & 1 & 0 & 0 \end{bmatrix}$$

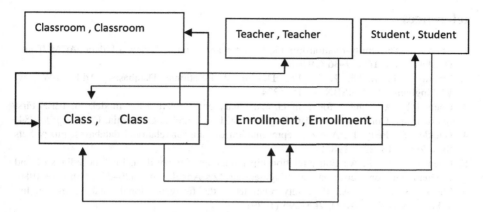

Fig. 19. The coherent graph

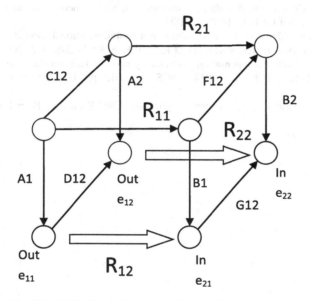

Fig. 20. Multi-dimension space reference (e_{11}, e_{12}, e_{21}, e_{22})

Now with A and B we change the "out" state and "in" state and generate the new relationship R_{12}. In Fig. 6, we have the principal relation R_{11} as the father of the all the other relations. The father generates two children, one is R_{12} and the other is R_{21}. The children R_{12} and R_{21} join to generate R_{22}. This is the morphogenetic process. Figure 6 is an aggregation of different databases with different states not only two of "out" and "in". The building of the new meta database gives us the instrument to move dynamically from one database to another or to separate one big database into more simple database one connected with the other by homotopic transformation or isomorphic transformation with deformation.

References

1. Chen, P.: The entity-relationship model - toward a unified view of data. ACM Trans. Database Syst. 1(1), 9–36 (1976)
2. Mannila, H., Räihä, K.-J.: The Design of Relational Databases. Addison-Wesley, Wokingham (1992). ISBN 0201565234
3. Chen, P.P.: An algebra for a directional binary entity-relationship model. In: IEEE First International Conference on Data Engineering, Los Angeles, California, USA, April 1984)
4. Buckles, B., Petry, F.: A fuzzy representation of data for relational databases. Fuzzy Sets Syst. 7(3), 213–226 (1982)
5. Chen, P.P., Zvieli, A.: Entity-relationship modeling of fuzzy data. In: Proceedings of 2nd International Conference on Data Engineering, Los Angeles, pp. 320–327, February 1986
6. Shenoi, S., Melton, A.: Proximity relations in the fuzzy relational database model. Int. J. Fuzzy Sets Syst. 31(3), 285–296 (1989)
7. Yager, R.R.: Concept representation and database structures in fuzzy social relational networks. IEEE Trans. Syst. Man, Cybern. Part A 40(2), 413–419 (2010)
8. Torlone, R.: Conceptual Multidimensional Models, Multidimensional Databases: Problems and Solutions. IGI Global, Hershey (2003)
9. Malinowski, E., Zimányi, E.: Hierarchies in a multidimensional model: from conceptual modeling to logical representation. Data Knowl. Eng. 59(2), 348–377 (2006)
10. Resconi, G.: Conflict compensation, redundancy and similarity in databases federation. In: Nguyen, N.T. (ed.) TCCI XIV 2014. LNCS, vol. 8615, pp. 120–135. Springer, Heidelberg (2014)
11. Wymore, W.: Model-Based Systems Engineering. CRC Press Inc., Boca Raton (1993)

A Method for Size and Shape Estimation in Visual Inspection for Grain Quality Control in the Rice Identification Collaborative Environment Multi-agent System

Marcin Hernes[1]([⊠]), Marcin Maleszka[2], Ngoc Thanh Nguyen[2], and Andrzej Bytniewski[1]

[1] Wrocław University of Economics, Wrocław, Poland
{marcin.hernes,andrzej.bytniewski}@ue.wroc.pl
[2] Faculty of Computer Science and Management,
Wrocław University of Technology, Wrocław, Poland
{marcin.maleszka,ngoc-thanh.nguyen}@pwr.edu.pl

Abstract. Computer vision methods have so far been applied in almost every area of our lives. They are used in medical sciences, natural sciences, engineering, etc. Computer vision methods have already been used in studies on the search for links between the quality of raw food technology and their external characteristics (e.g. color, size, texture). Such work is also conducted for cereals. For the analysis results to meet the expectations of users of the system, it should include not only the attributes describing the controlled products, materials or raw materials, but should also indicate the type of material or species/variety of raw material. However existing solutions are very often implemented as closed source software (black box) therefore the user has no possibility to customize them (for example the enterprise cannot integrate these solutions into its management information system). The high cost of automated visual inspection systems are also a major problem for enterprises. The aim of this paper is to develop a method of estimating the size and shape of a rice grains using visual quality analysis, implemented in the multi-agent system named Rice Identification Collaborative Environment. Using this method will allow statistical analysis of the characteristics of the sample, and will be one of the factors leading to the identification of species/varieties of cereals and determining the percentage of the grains that do not meet quality standards. The method will be implemented as an open source software in Java. Consequently it can be easily integrated into enterprise's management information system. Because it will be available for free, the cost of automated visual inspection systems will be reduced significantly. This paper is organized as follows: the first part shortly presents the state-of-the-art in the considered field; next, a developed method for size and shape estimation implemented in the Rice Identification Collaborative Environment is characterized; the results of a research experiment for verification of the developed method are presented in the last part of paper .

Keywords: Automated visual inspection · Image recognition · Grain quality control · Multi-agent systems

© Springer-Verlag Berlin Heidelberg 2016
N.T. Nguyen and R. Kowalczyk (Eds.): TCCI XXII, LNCS 9655, pp. 205–217, 2016.
DOI: 10.1007/978-3-662-49619-0_11

1 Introduction

The quality control on manufacturing lines is often done automatically by using non-contact visual (vision, optical) methods (DIA - digital image analysis) also called an automated visual inspection [12]. Their advantages are high efficiency, high performance and no need for operator intervention [15]. Visual quality control systems are usually equipped with a camera and specialized software for image processing and analysis. This software enables identification of information significant from the point of view of quality control, contained in the picture. On the basis of this information the qualitative selection of products is performed [5]. There are three groups of vision system: vision sensors, compact vision systems and vision systems implemented using PCs [13]. Individual groups are divided in terms of functionality, technical characteristics, prices, possible applications.

Computer vision methods have already been used in studies on the search for links between the quality of raw food technology and their external characteristics (attributes, e.g. color, size, texture). Such works are also conducted for rice grains. The relationship between the dimensions of grains, seed coat color, surface texture and gluten content, and rheological properties are considered [2]. However, for the analysis results to meet the expectations of users of the system, it should include not only the attributes describing the controlled products, materials or raw materials, but should also indicate the type of material or species/ variety of raw material.

However, existing solutions are very often implemented as a closed source software (blackbox) therefore the user has no possibility to customize them (for example the enterprise cannot integrate these solutions into its management information system). The high cost of automated visual inspection systems are also a major problem for enterprises.

The aim of this paper is to develop a method of estimating the size and shape of grain cereals using visual quality analysis, implemented in the multi-agent system named Rice Identification Collaborative Environment (RICE). Using this method will allow statistical analysis of the characteristics of the sample, and will be one of the factors leading to the identification of species/varieties of cereals and determining the percentage of the grains that do not meet quality standards. The method will be implemented as open source software in Java. Consequently it can be easy integrated into enterprise's management information system. Because it will be available for free, the cost of automated visual inspection systems will be reduced significantly.

Our wider research goal is the development of a fully functional automated visual inspection system for rice quality control. This system is developed for the needs of several Polish enterprises manufacturing rice products. This paper presents preliminary results of this research and it is organized as follows: the first part shortly presents the state-of-the-art in the considered field; next, the developed method for size and shape estimation implemented in the Rice Identification Collaborative Environment is characterized; the results of a research experiment for verification of the developed method are presented in the last part of paper.

2 Related Works

Computer vision methods (diagnostic digital image analysis) have been so far applied in almost every area of our life. They are used in medical sciences, natural sciences, engineering, etc. Vision systems are used, inter alia, to control various processes, identify characteristics of objects. In the last decade there was a large increase in the number of research projects related to assessing the quality of food and agricultural products on the basis of objective instrumental measurements, especially techniques based on image recognition. Using this kind of techniques can be used not only to analyze, evaluate, but also to classify individual product features, such as color, texture, shape, size, and to define the relationship between these parameters [17]. Computer image analysis is a nondestructive method that allows obtaining fast, reproducible and objective evaluation of the quality. It is increasingly used for measuring and predicting the quality of agricultural and food commodities, while sometimes it overcomes the limitations of traditional methods used so far.

Many research centers focus on exploring the relationships between technological quality of food raw materials and their external characteristics, identifiable by the use of vision systems (color measurement, geometry, surface textures). Such work is also carried out for cereals. The relationship between the dimensions of grain, seed coat color, surface texture and gluten content, and rheological properties are the most often used features of grain. Many authors used a technique of digital recording to identify the various types of cereals [14]. One approach is based on a flatbed scanner to identify different varieties of Indian wheat. Another approach was to develop a system to distinguish 31 varieties of wheat using a CCD camera.

More and more research on engineering issues and manufacturing technology concerns the possibility of using a fully automated and error-free continuous monitoring and quality control. In particular, the control area associated with the measurement of quality characteristics, including the physical properties measured on-line or off-line is the most desirable direction for development [1]. The research [4] shows that visual marbling assessment may be appropriate as an assessment on the perception of sensory properties of meat. However, some of the authors stated that the selection of raw material based solely on visual assessment of marbling cannot replace sensory profile data [4], although other authors suggest that there are already solutions that fully meet the expectations in this area [6].

Automated grain size and shape estimation methods, supported by unconsolidated sediment digital image processing, can be grouped into two trends [7]. The first is the statistical approach, which has been progressively adopted to estimate grain size for images of small sediment fractions, such as sand. This technique is based on the principle that the degree to which nearby pixels are correlated on an image varies with grain size. That is, nearby pixels on an image of coarse grains are more correlated (i.e., their patterns have a high probability of being similar) than in samples of finer grains [14]. This method, was improved by [3] through a bi-dimensional Fourier transform to avoid the use of a specific database of standardized images (previously analyzed) of the study site, usually known as the beach catalogue (which is used to characterize new images through correlation). This particular method has the disadvantage of working only with images acquired in controlled environments and is dependent on the hardware specifications. One of the main limitations of these statistical approaches is that only the average grain size is estimated from the entire image, and characteristics of individual grains are not exploited, thus it is not able to derive the complete grain size distribution.

The second technique, known as the image segmentation method, has been extensively used in biomedical applications and was adapted to grain size and shape estimation [9]. The goal of this method is to detect the grain boundaries in an automated process allowing the estimation of the grain size distribution over discrete grain size classes. The main difficulty of detecting the grain boundaries in this method is the absence of a contrasting image to isolate the pixel boundaries of the individual particles. In fact, the full delineation of each grain in the image is probably the crucial part in automatic grain identification, a task that is more difficult when the image resolution is reduced and when grain overlapping is more frequent. Mostly for this reason, previous implementations of this methodology have been essentially used to estimate larger sediment fractions, such as coarse gravels or surfaces composed of distinctive bimodal mixtures that include gravel fractions [9]. In sand samples, grains are reasonably homogeneous in size, causing frequent overlapping and poor visual delineation of the grains, which added to the resulting small image resolution (i.e., less pixels per grain, in average) when compared to gravel size samples, may compromise the results of such automated image processing algorithms. Moreover, the method usually chosen for fully delineating the image grains is the watershed. This method may result in over-sectioning the grains in the image, jeopardizing the correct estimation of the grain dimensions.

However, the present image based methodologies for the analysis of grain characteristics, rely on controlled image-capturing conditions and on specific hardware structures that are customized for this application, like:

- a waterproof housing with an LCD light ring to provide an evenly illuminated sediment bed [16],
- a black box with the camera in a housing mounted at the top to guarantee images at a constant height above the surface [3],
- a tripod to fix the camera orthogonal to the observed surface [19].

The paper [2] proposed size estimation process divided into three main stages:

(1) the image acquisition procedure;
(2) image processing analysis with automated grain size estimation;
(3) user validation of grain identification results.

This method was designed to allow a simple and efficient data acquisition procedure without the need for dedicated equipment. The image processing analysis is supported by a search for variations in the gradient of the pixel intensity. This approach is suitable for processing sand-sized grain samples, in contrast with the target samples of other proposed algorithms that consider gravel and cobble-sized samples.

The computer vision methods presented in this section are very often implemented as closed source software (blackbox) and their cost for enterprises is high. In addition, these methods are implemented in different programming languages, therefore it is very difficult to integrate them consistently into management information system.

Therefore it is necessary to develop the method which will be implemented as open source software and it will be easily integrated into enterprise's management information system. Such method is presented in the next part of this paper. For the test purpose, the method is implemented in the prototype of RICE system.

3 A Method for Size and Shape Estimation in the Rice Identification Collaborative Environment

In our research we develop an open modular system for rice identification. The final application will be an open-source system available online. For purposes of this paper the base structure of modules was developed, as well as some basic identification algorithms. In general, the system input is the image of multiple rice grains (on black background) and the output is the statistics of this rice.

The Rice Identification Collaborative Environment (RICE) is a multi-agent system based on The Learning Intelligent Distribution Agent (LIDA), characterized in details in [8, 10]. In short words, in the LIDA architecture it was adopted that the majority of basic operations are performed by the so-called codelets, namely specialized, mobile programs processing information in the model of global workspace. The functioning of the cognitive agent is performed within the framework of the cognitive cycle and it is divided into three phases: the understanding phase, the consciousness phase and the selection of actions and learning phase. At the beginning of the understanding phase the stimuli received from the environment activate the codelets of the low level features in the sensory memory [8]. The outlets of these codelets activate the perceptual memory, where high level feature codelets supply more abstract things such as objects, categories, actions or events. The perception results are transferred to the workspace and on the basis of episodic and declarative memory local links are created. Then, with the use of the occurrences of perceptual memory, a current situational model is generated; it other words the agent understands what phenomena are occurring in the environment of the organization. The consciousness phase starts with forming of the coalition of the most significant elements of the situational model, which then compete for attention so the place in the workspace, by using attentional codelets. The contents of the workspace module are then transferred to the global workspace (the so-called "broadcasting" takes place), simultaneously initializing the phase of action selection. At this phase possible action schemes are taken from procedural memory and sent to the action selection module, where they compete for the selection in a given cycle. The selected actions activate sensory-motor memory for the purpose of creating an appropriate algorithm of their performance, which is the final stage of the cognitive cycle [11]. The cognitive cycle is repeated with the frequency of 5–10 times per second.

Figure 1 presents the functional architecture of RICE.

The two main groups of agents are the singular Management Agent and the Identification Agents.

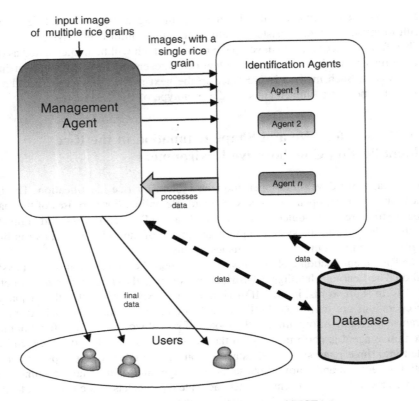

Fig. 1. The functional architecture of RICE.

The Management Agent is tasked with dividing the input image of multiple rice grains into a number of images, each with a single rice grain only. It also gathers and processes data generated by Identification Agents and presents the final data to the user. Multiple methods may be used to divide the image. For the prototype we created a basic contour tracking algorithm (move through image until non-black pixel, then follow contour; cut out the box including the contour), but we did not use it in experiments due to environmental factors (it requires precise lighting conditions for taking images).

The Identification Agents are the main element of the rice identification system. Multiple agents may be used and they may each use a different method of rice grain identification. In the most basic application they may also use the same method, to speed up processing by parallel computation. For the prototype we have created two methods to be used by the agents, used to identify:

- Average color saturation – this very basic method works by determining the ratio of each base color in the whole image (i.e. the number of pixels with Red/Green/Blue over a given threshold to the total number of pixels; note that a pixel may be counted in more than one color). This may be used to identify rice grain by its average color (white, brown, etc.).

- Main diagonals – this method is used to determine length, width and two additional diagonals (lower left to upper right and upper left to lower right). First the center of the rice grain is determined (by counting the *weight* of non-black pixels). Then the longest diagonal is determined by checking all outermost non-black pixels (with additional checking if the line to the center is also non-black). With length determined, the width is calculated as the longest possible line perpendicular to the length. Two other lines are checked additionally, but they are not necessary for determining rice grain size and shape. They are calculated by finding the longest line close to bisector of the angle given by length and width. The algorithm of this method is presented on Fig. 2.

The data generated by all agents is stored in a database.
The implementation of RICE is realized as follows:

1. Communication architecture between agent modules was ensured by using LIDA framework's codelets.
2. Communication between agents is based on Java Message Service (JMS) technology. The representation of data (generated in result of agents' operating) in form of XML format document, was adopted (the JMS messaging is at the text type). The communication is realized in publish/subscribe messaging domains – it guarantees that information or knowledge generated by one of agents is immediately available for the other agents. The asynchronous message consumption is used.

All of the agents' functions are available as a local services or e-services by using Web Services technology.

At the physical level, the RICE is built on the basis of two main technologies – the LIDA framework (due to the framework being developed in Java language and as it is open to the implementation of the other Java technologies – mentioned JMS, Java Database Connectivity or Java API for XML Web Services) and Microsoft SQL Server 2008 database management system.

Next section of this paper presents results of research experiment performed in order to evaluate the developed method.

In the integration process of different technologies some inconsistency can appear and consensus methods can be very useful for solving [18, 19].

4   Research Experiment

In order to verify the developed method, a research experiment has been carried out, in which the following assumptions were made:

1. Experiment concerns the main diagonals method.
2. The digital camera was held in a tripod to guarantee that the images were taken at a constant height above the surface, and the focal plane of the camera was maintained parallel to the sediment surface.

```
Input: Bitmap with a single rice grain
Output: Length, Width
Parameter: Color threshold τ

1. Sum_pixels_X = 0, Sum_pixels_Y = 0, Count_Pixels = 0
2. For each pixel P with coordinates X and Y
2.1. If color of P > τ
2.1.1. Increase Sum_pixels_X by X
       Increase Sum_pixels_Y by Y
       Increase Count_Pixels by 1
3. Center_of_rice_X = Sum_pixels_X / Count_Pixels
   Center_of_rice_Y = Sum_pixels_Y / Count_Pixels

4. a = 0, b = 0, Length = 0
5. For each pixel P with coordinates X and Y
5.1. If color of P > τ
5.1.1. a = (Y - Center_of_rice_Y) /
           / (X - Center_of_rice_X)
       b = Y - a * X
5.1.2. For each pixel I lying on line y = a*x + b in
       range of x = Center_of_rice_X to X
5.1.2.1. If color of I < τ goto 5.
5.1.3. Temp_length = 2 * distance between P and
       (Center_of_rice_X,Center_of_rice_Y)
5.1.4. If Temp_length > Length
5.1.4.1. Length = Temp_length
6. Width = 0
7. For each pixel P with coordinates X and Y
7.1. If color of P > τ
7.1.1. a1 = -1 / a
       b1 = Y - a1 * w
       pX = (b1 - b) / (a - a1)
       pY = a1 * pX  + b1
7.1.2. For each pixel I lying on line y = a1*x + b1 in
       range of x = X and pX
7.1.2.1. If color of I < τ goto 7.
7.1.3. Temp_width = 2 * distance between P and (pX,pY)
7.1.4. If Temp_width > Width
7.1.4.1. Width = Temp_width
```

Fig. 2. The algorithm of main diagonals method

3. All of the images were taken in natural sunlight conditions, with the same illumi-
   nation angle, although the surface plane of the sediments was sheltered from direct
   sunlight (therefore average color saturation method is not tested in this experiment -
   it must be tested in artificial, fixed light condition).
4. Black background has been used.
5. Three rice varieties have been tested:

   - white rice,
   - natural rice,
   - wild rice.

6. Relative measure will be used – the grains' size is presented in pixels.
7. The number of grains have been selected randomly (range is 20–50 grains).
8. The following measures have been used: length/width proportion, average, standard
   deviation and dominant.
9. The length/width proportion is used also for shape estimation.

The results of white rice size estimation are presented in Table 1.

Table 1. The results of white rice size estimation.

Grain number	Length	Width	Length/width proportion	Grain number	Length	Width	Length/width proportion
1	182	66	2,758	12	182	72	2,528
2	168	70	2,400	13	148	68	2,176
3	160	58	2,759	14	148	62	2,387
4	190	68	2,794	15	178	74	2,405
5	156	50	3,120	16	170	72	2,361
6	174	76	2,289	17	194	66	2,939
7	169	74	2,284	18	160	58	2,759
8	192	68	2,824	19	190	68	2,794
9	168	70	2,400	20	80	60	1,333
10	156	50	3,120	21	84	74	1,135
11	170	78	2,179	22	76	62	1,226
				Average	**158,864**	**66,545**	**2,408**
				Std dev.	**33,232**	**7,394**	**0,530**
				Dominant	**182**	**68**	**2**

The white rice is characterized by the average length/width proportion equal to
2,408 and the dominant of this measure is 2. The grains no. 20–22 are damaged,
therefore length/width proportion is near 1.

The results of natural rice size estimation are presented in Table 2.

The natural rice is characterized by the average length/width proportion equal to
2,126 and the dominant of this measure is 2. The grains no. 40–50 are damaged,
therefore length/width proportion is lower than in the case of undamaged grains.

Table 2. The results of natural rice size estimation

Grain number	Length	Width	Length/width proportion	Grain number	Length	Width	Length/width proportion
1	108	48	2,250	26	90	40	2,250
2	106	38	2,789	27	126	52	2,423
3	112	36	3,111	28	120	114	1,053
4	130	58	2,241	29	126	40	3,150
5	116	44	2,636	30	122	48	2,542
6	110	36	3,056	31	116	42	2,762
7	102	58	1,759	32	126	38	3,316
8	122	40	3,050	33	114	40	2,850
9	122	40	3,050	34	102	46	2,217
10	104	100	1,040	35	112	42	2,667
11	122	40	3,050	36	120	38	3,158
12	104	46	2,261	37	112	66	1,697
13	114	40	2,850	38	124	34	3,647
14	126	38	3,316	39	108	50	2,160
15	126	40	3,150	40	64	58	1,103
16	118	34	3,471	41	74	36	2,056
17	90	40	2,250	42	78	40	1,950
18	126	52	2,423	43	70	44	1,591
19	124	50	2,480	44	54	52	1,038
20	116	42	2,762	45	58	40	1,450
21	118	34	3,471	46	72	42	1,714
22	126	38	3,316	47	70	44	1,591
23	114	40	2,850	48	54	52	1,038
24	124	50	2,480	49	56	46	1,217
25	126	40	3,150	50	74	40	1,850
				Average	91,182	44,455	2,126
				Std dev.	26,008	7,197	0,751
				Dominant	126	40	2

The results of wild rice size estimation are presented in Table 3.

The wild rice is characterized by the average length/width proportion equal to 4,741 and the dominant of this measure is 4. The grains no. 30–32 are damaged, therefore length/width proportion is lower than in the case of undamaged grains.

Taking into account all three varieties of rice, it can be noted that they are characterized by different average length and width. The wild rice is characterized by highest value of length/width proportion, standard deviation of this measure and its length value is highest. However, the average width and width standard deviation values are lowest in case of wild rice. Higher values of these measures are in case of white rice.

Table 3. The results of wild rice size estimation

Grain number	Length	Width	Length/width proportion	Grain number	Length	Width	Length/width proportion
1	218	44	4,955	17	187	41	4,561
2	232	42	5,524	18	206	32	6,438
3	234	44	5,318	19	202	42	4,810
4	314	50	6,280	20	211	38	5,553
5	242	48	5,042	21	224	42	5,333
6	238	43	5,535	22	220	34	6,471
7	178	34	5,235	23	168	41	4,098
8	196	43	4,558	24	184	43	4,279
9	212	40	5,300	25	216	42	5,143
10	217	44	4,932	26	228	46	4,957
11	144	36	4,000	27	186	34	5,471
12	168	41	4,098	28	246	50	4,920
13	154	35	4,400	29	234	44	5,318
14	185	32	5,781	30	114	34	3,353
15	178	41	4,341	31	103	40	2,575
16	195	37	5,270	32	116	37	3,135
				Average	**184,955**	**39,182**	**4,741**
				Std dev.	**37,870**	**4,478**	**0,943**
				Dominant	**168**	**41**	**4**

The length/width proportion may be also used for grain shape estimation, as this measure allows for determining damaged grains.

On the basis of experimental results it can be also concluded that the elaborated method can be used as an element for determining the grain varieties. However, more parameters are needed in this purposes, mainly the color determination. This method allow also for determining damaged grains which do not meet quality standards. The length/width proportion is an indicator of such grains.

5 Conclusions

The estimation of the size of individual grains is a very important element of visual inspection used in quality control process. On the basic of experimental results it can be stated that the method developed in this paper allows the measurement of the size of grains and consequently it can be used in grain quality inspection. The developed method is one of the factors leading to the identification of species/varieties of cereals and determining the percentage of the grains that do not meet quality standards. The method is implemented in Java as open source software, therefore it can be easily integrated into enterprise's management information system and consequently, the cost of visual quality control system is lower for enterprises.

The further research may concern, among other, developing the method for color recognizing. For this process some existing methods for fruit defect detection can be used [20]. Also methods for path planning for autonomous vehicle can be useful because the grains are in moving [21].

References

1. Abdullah, M.Z., Guan, L.C., Lim, K.C., Karim, A.A.: The applications of computer vision system and tomographic radar imaging for assessing physical properties of food. J. Food Eng. **61**, 125–135 (2004)
2. Baptista, P., Cunha, T.R., Gama, C., Bernardes, C.: A new and practical method to obtain grain size measurements in sandy shores based on digital image acquisition and processing. Sed. Geol. **282**, 294–306 (2012)
3. Buscombe, D., Masselink, G.: Grain-size information from the statistical properties of digital images of sediment. Sedimentology **56**, 421–438 (2009)
4. Cannata, S., Engle, T.E., Moeller, S.J., Zerby, H.N., Radunz, A.E., Green, M.D., Bass, P.D., Belk, K.E.: Effect of visual marbling on sensory properties and quality traits of pork loin. Meat Sci. **85**, 428–434 (2010)
5. Fornal, J., Jeliński, T., Sadowska, J., Quattrucci, E.: Comparison of endosperm microstructure of wheat and durum wheat using digital image analysis. Int. Agrophys. **13**(2), 215–220 (1999)
6. Fortin, A., Robertson, W.M., Tong, A.K.W.: The eating quality of Canadian pork and its relationship with intramuscular fat. Meat Sci. **69**, 297–305 (2005)
7. Fornal, Ł., Kozirok, W., Chorazewicz, R.: New possibilities to characterizing wheat grain endosperm. Pol. J. Food Nutr. Sci. **13**(1), 170–183 (2003)
8. Franklin, S., Patterson F.G.: The LIDA architecture: adding new modes of learning to an intelligent, autonomous, software agent. In: Proceedings of the International Conference on Integrated Design and Process Technology. Society for Design and Process Science, San Diego (2006)
9. Graham, D.J., Reid, I., Rice, S.P.: Automated sizing of coarse-grained sediments: image-processing procedures. Math. Geol. **37**, 1–28 (2005)
10. Hernes, M., Sobieska-Karpińska, J.: Application of the consensus method in a multiagent financial decision support system. Inf. Syst. e-Bus. Manag., 1–19 (2015). doi:10.1007/s10257-015-0280-9. Springer, Heidelberg
11. Hernes, M., Maleszka, M., Nguyen, N.T., Bytniewski, A.: The automatic summarization of text documents in the Cognitive Integrated Management Information System. In: Proceedings of Federated Conference Computer Science and Information Systems (FedCSIS), Łódź (2015)
12. Iqbal, A., Valous, N.A., Mendoza, F., Sun, D.-W., Allen, P.: Classification of pre-sliced pork and Turkey ham qualities based on image colour and textural features and their relationships with consumer responses. Meat Sci. **84**, 455–465 (2010)
13. Majumdar, S., Jayas, D.S.: Classification of cereal grains using machine vision: VI. Combined morphology, color, and texture models. Am. Soc. Agric. Eng. **43**(6), 1689–1694 (2000)
14. Rubin, D.M., Chezar, H., Harney, J.N., Topping, D.J., Melis, T.S., Sherwood, C.R.: Underwater microscope for measuring spatial and temporal changes in bed-sediment grain size. Sed. Geol. **202**, 402–408 (2007)

15. Sànchez, A.J., Albarracin, W., Grau, R., Ricolfe, C., Barat, J.M.: Control of ham salting by using image segmentation. Food Control **19**, 135–142 (2008)
16. Visen, N.S., Paliwal, J., Jayas, D.S., White, N.D.G.: Specialist neural networksfor cereal grain classification. Biosyst. Eng. **82**(2), 151–159 (2001)
17. Warrick, J.A., Rubin, D.M., Ruggiero, P., Harney, J.N., Draut, A.E., Buscombe, D.: Cobble cam: grain-size measurements of sand to boulder from digital photographs and autocorrelation analyses. Earth Surf. Proc. Land. **34**, 1811–1821 (2009)
18. Hernes, M., Nguyen, N.T.: Deriving consensus for hierarchical incomplete ordered partitions and coverings. J. Univ. Comput. Sci. **13**(2), 317–328 (2007)
19. Nguyen, N.T.: Consensus systems for conflict solving in distributed systems. Inf. Sci. **147** (1–4), 91–122 (2002)
20. Pham, V.H., Lee, B.R.: An image segmentation approach for fruit defect detection using k-Means clustering and graph-based algorithm. Vietnam J. Comput. Sci. **2**(1), 25–33 (2015)
21. Hoang, V.D., Jo, K.H.: Path planning for autonomous vehicle based on heuristic searching using online images. Vietnam J. Comput. Sci. **2**(2), 109–120 (2015)

Author Index

Printed in the United States
By Bookmasters